Isma Younes
Muhammad Shafiq
Abdul Ghaffar
Shahid Mehmood

Spatial Patterns of Noise Pollution and its Effects in Lahore City

Anchor Academic
Publishing

Younes, Isma, Shafiq, Muhammad, Ghaffar, Abdul, Mehmood, Shahid: Spatial Patterns
of Noise Pollution and its Effects in Lahore City, Hamburg, Anchor Academic
Publishing 2017

Buch-ISBN: 978-3-96067-139-8
PDF-eBook-ISBN: 978-3-96067-639-3
Druck/Herstellung: Anchor Academic Publishing, Hamburg, 2017

Bibliografische Information der Deutschen Nationalbibliothek:
Die Deutsche Nationalbibliothek verzeichnet diese Publikation in der Deutschen
Nationalbibliografie; detaillierte bibliografische Daten sind im Internet über
http://dnb.d-nb.de abrufbar.

Bibliographical Information of the German National Library:
The German National Library lists this publication in the German National Bibliography.
Detailed bibliographic data can be found at: http://dnb.d-nb.de

All rights reserved. This publication may not be reproduced, stored in a retrieval system
or transmitted, in any form or by any means, electronic, mechanical, photocopying,
recording or otherwise, without the prior permission of the publishers.

Das Werk einschließlich aller seiner Teile ist urheberrechtlich geschützt. Jede Verwertung
außerhalb der Grenzen des Urheberrechtsgesetzes ist ohne Zustimmung des Verlages
unzulässig und strafbar. Dies gilt insbesondere für Vervielfältigungen, Übersetzungen,
Mikroverfilmungen und die Einspeicherung und Bearbeitung in elektronischen Systemen.

Die Wiedergabe von Gebrauchsnamen, Handelsnamen, Warenbezeichnungen usw. in
diesem Werk berechtigt auch ohne besondere Kennzeichnung nicht zu der Annahme,
dass solche Namen im Sinne der Warenzeichen- und Markenschutz-Gesetzgebung als frei
zu betrachten wären und daher von jedermann benutzt werden dürften.

Die Informationen in diesem Werk wurden mit Sorgfalt erarbeitet. Dennoch können
Fehler nicht vollständig ausgeschlossen werden und die Diplomica Verlag GmbH, die
Autoren oder Übersetzer übernehmen keine juristische Verantwortung oder irgendeine
Haftung für evtl. verbliebene fehlerhafte Angaben und deren Folgen.

Alle Rechte vorbehalten

© Anchor Academic Publishing, Imprint der Diplomica Verlag GmbH
Hermannstal 119k, 22119 Hamburg
http://www.diplomica-verlag.de, Hamburg 2017
Printed in Germany

Dedicated to our families

Preface

Lahore is one of the worst effected cities due to uncontrolled noise pollution in Pakistan. The most important factor of noise pollution is the road traffic. The main objective of this study was to analyze and evaluate road traffic noise and to see its effects on the population of Lahore city. A weighting sound level meter was used in the study. All the measurements were taken at a height of about 1.2 m from the ground. Noise measurements were taken at fifty six sample sites. Spatial pattern of noise was shown in the maps. Maps were also drawn to show buffers dividing areas into moderate, high and extremely high risk zones in accordance with noise risk levels. Day and night time noise maps were also drawn with the graduated symbol. The mean day-night values were exceeding the permissible environmental standards used in Pakistan. A survey was done to study the diseases caused by noise pollution in the areas with highest noise levels. It was found that 100% of respondents were suffering from temporary hearing loss.

Contents

Preface i

List of abbreviations iv

Contents: iii

Chapter 1: Introduction 1
    1.1   Background...3
    1.2   Aims and objectives of the study 5
    1.3   Study area: 5
    1.4   Topography...7
    1.5   Temperature..7
    1.6   Rainfall 8
    1.7   Wind speed and direction 9
    1.8   The road network...10
    1.9   Status of vehicular traffic in Lahore...12
    1.10   Impacts of noise pollution on urban life..13
        1.10.1 Audiological impacts 13
        1.10.2 Biological impacts 13
        1.10.3 Behavioral impacts..13

Chapter 2: Literature review 15
    2.1   Literature review 15

Chapter 3: Noise pollution and its effects 27
    3.1   Introduction 27
    3.2   Measurement of noise 28
    3.3   Unit and range of noise 28
    3.4   Sources of noise pollution 30
        3.4.1   Point source..30
        3.4.2   Line source...30
        3.4.3   Plane source. 31
        3.4.4   Industrial source...31
        3.4.5   Non-industrial source...31
    3.5   Mechanism of Hearing...31
    3.6   Effects of noise...33
    3.7   Environmental Quality Standards (EQS) for noise in Japan...............36
    3.8   The duration of exposure to noise in USA..37
    3.9   National Environmental Quality Standards (NEQS) for Noise 38

Chapter 4: Data sources and methodology 39
    4.1   Introduction 39
    4.2   Data Collection 42
        4.2.1   Data Sources 42
        4.2.2   Equipment used for noise measurement 42
        4.2.3   Measurement Procedure 43
    4.3   Methodology 44
        4.3.1   Site selection 44
        4.3.2   Sample location 44
    4.4   Use of multivariate analysis techniques 46
        4.4.1   Cluster analysis and Factor analysis 46
    4.5   Presentation 46

Chapter 5: Spatial patterns of noise pollution 47
    5.1   Introduction 47
    5.2   Level of Noise Pollution around Walled City Lahore, 2011 53
    5.3   Level of Noise Pollution around Mall Road Lahore, 2011 56
    5.4   Level of Noise Pollution around Jail Road Lahore, 2011 59
    5.5   Level of Noise Pollution around Garhi Shahu Lahore, 2011 61
    5.6   Area wise Coverage of Noise Pollution 64
    5.7   Comparative analysis of morning and evening levels 66
    5.8   Statistical Analysis 72
        5.8.1   Cluster analysis 76
        5.8.2   Factor analysis 79

Chapter 6: Summary, conclusion, recommendation & future work 83
    6.1   Summary 83
    6.2   Conclusion 84
    6.3   Recommendations 87
    6.4   Scope for Future Work 88

Annex-1: Bibliography 89

Annex-2: NEQS for noise pollution 94

Annex-3: Questionnaire: Causes and effects of noise pollution 95

Annex-4: Noise levels at seventy six sample sites 97

Annex-5: Communalities 99

Annex-6: Component matrix(a) 101

Annex-7: Rotated Component Matrix(a) 103

Annex-8: Component Transformation Matrix 105

List of abbreviations

EPA	Environment Protection Agency
GIS	Geographic Information System
GPO	General Post Office
GPS	Global Positioning System
HL	Hearing Loss
LDA	Lahore Development Authority
LMA	Lahore Metropolitan Area
NEQS	National Environmental Quality Standards
NIHL	Noise Induced Hearing Loss
NITTS	Noise Induced Temporary Threshold Shift
NPL	Noise Pressure Level
SPL	Sound Pressure Level
WHO	World Health Organization

Chapter 1:
Introduction

Environmental pollution such as noise air, water and solid waste pollution has always been a global concern affecting the health of public. High concentration of environmental pollutants is increasing the threats to the quality of the environment. Air pollution, which is a release of chemicals and particles in to the atmosphere, causes many diseases like asthma, cancer and many bronchial problems. Water pollution that is addition of surface runoff, leakage into ground water, liquid spills, waste water discharges and addition of litter into water causes cholera, typhoid and many more diseases. Unlike all other pollutions causing components of environment, sound is not an element, compound or substance which can accumulate and harm future generations. Noise is a special feature of cities.

It is a special kind of wave action usually transmitted by air in form of pressure waves and received by hearing apparatus present in body of animals. It is a sound without value that is undesired by the recipient. Noise is playing an ever increasing role in our lives and seems a regrettable but ultimately avoidable result of the current technology. Noise measurements include subjective as well as objective factors. A physical measurement of noise magnitude must be supported with subjective loudness and annoyance related factors. The sound pressure level is an objective quantification of noise based on measured sound pressure. The effect of noise on humans however, depends not only on its magnitude but also on its frequency content because the ear is not equally sensitive to noise at all frequencies in the audible range of 20-20,000 Hz. Noise levels in general have increased over the years. The noise levels in the cities have increased at about 1 dB per year for the last 30 years (Wang, 2005).

The present study will focus on the patterns of noise pollution in Lahore City. Lahore is a typical inland city of Pakistan. As shown in Figure 1.1, this land locked city is situated about 1100 kilometres away from the Arabian Sea. Lahore is one of the worst affected cities due to unchecked noise pollution.

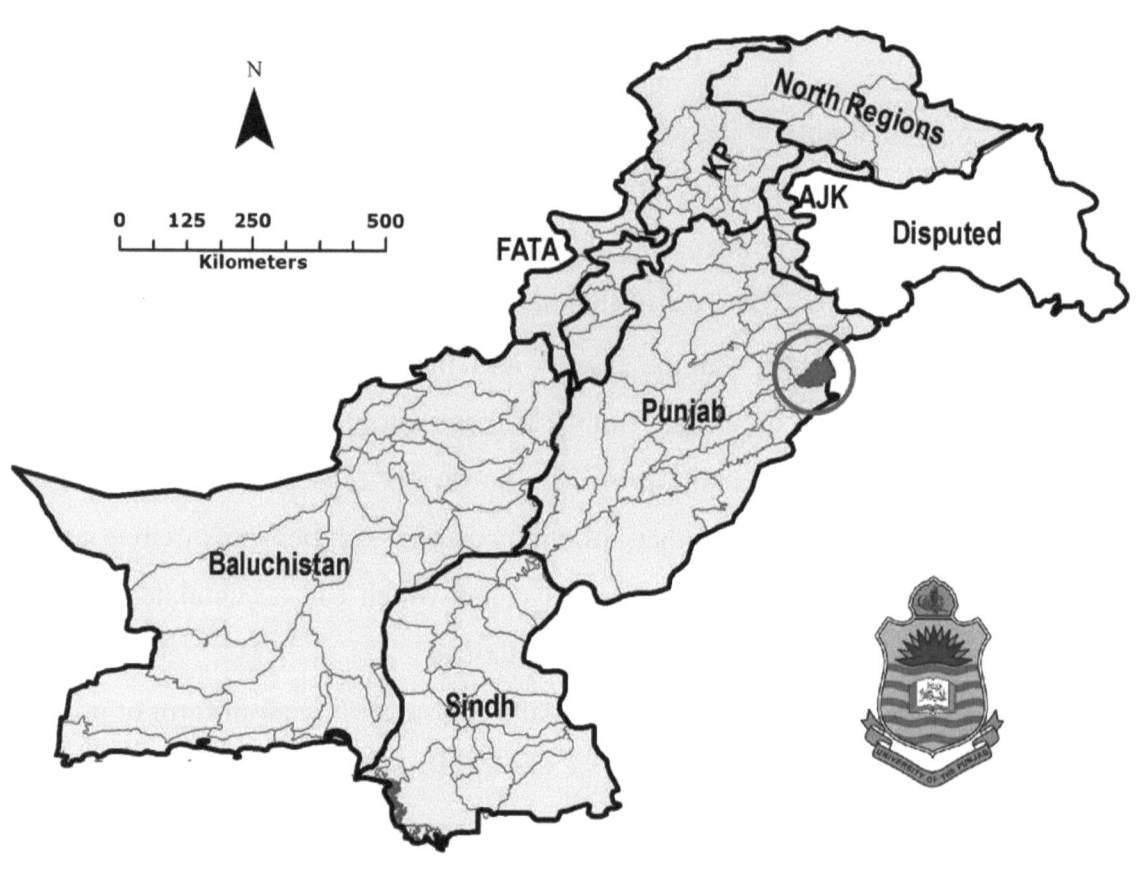

Figure 1.1: Location of the study area (Lahore City) is shown in red circle

The geographical location of Lahore is from 31° 13' 28'' N to 31° 43' 02'' N in latitude and 74° 00' 00'' E to 74° 39' 05'' N in longitude. Situated along the South bank of River Ravi, the city is bounded by Sheikhupura district in the North and Kasur District in the South. The east of Lahore city is the International Boundary

Line separating Pakistani Punjab from Indian Punjab. The adjoining city on the Indian side is Amritsar (Ajnala, Tarn Taran and Patti) as shown in Figure 1.2.

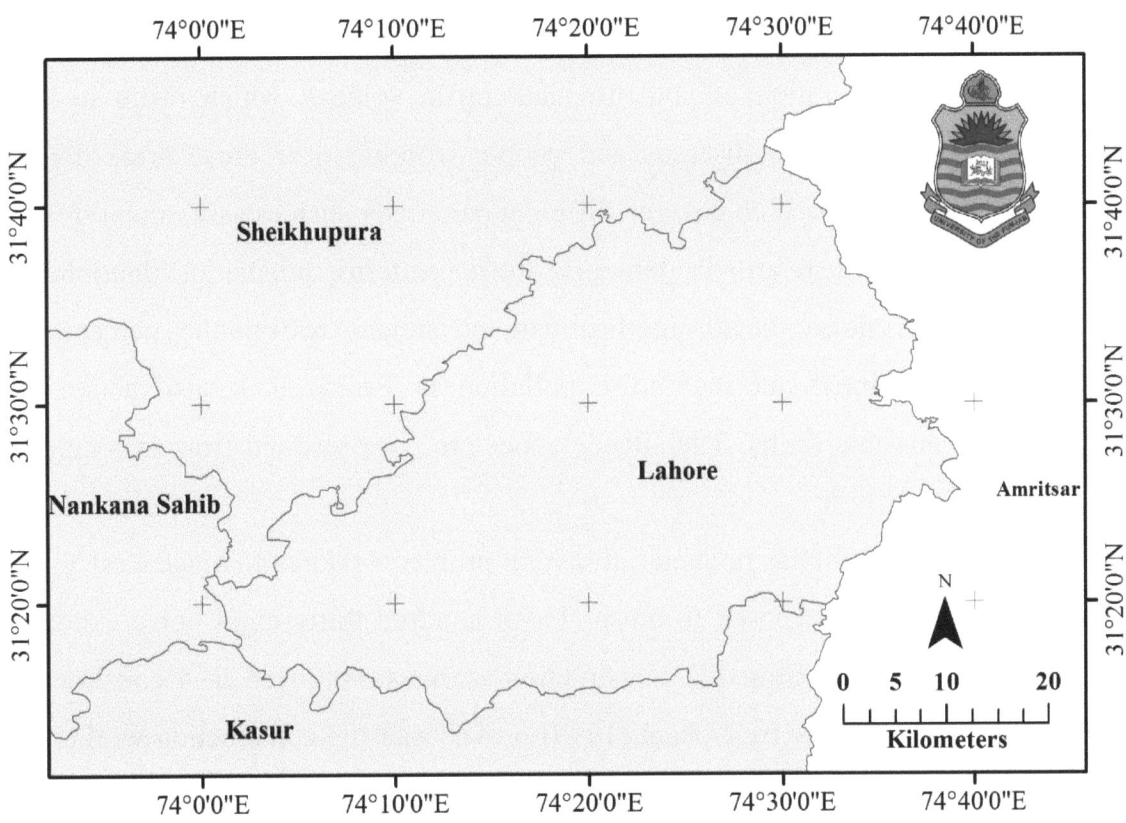

Figure 1.2: Geographical Location of Lahore City

There are several factors that contribute to increase of noise levels in the City. One of these factors is the increase in urban population that contributed to high traffic volume. Other factors include the limited space to live in the City, high rise buildings as well as high traffic volume, which gave birth to multiple problems related to noise.

1.1 Background

The recognition of traffic noise as one of the major cause of environmental pollution has led to the research in this field. The effect of noise on human health has put many questions in the minds of the people. The effects of noise pollution on human beings have been intensively carried out. Many researchers carried out their research in this field. The spatio-temporal distribution of noise was studied in West Bengal generated by urban traffic, by means of monitoring and mapping as a tool for evaluation of impacts (Banerjee, 2009). Mehdi (2002) concluded in a research in Karachi on noise pollution that the high level of noise is associated with the geographical agglomeration of land use and traffic volume, which result in high incidence of noise related diseases and people working near those areas are on vulnerable risks. The research covered different parameter such as assessment of land cover, human settlement growth, temporal traffic patterns, population distribution, current levels of noise, health implications, physicians and public perceptions. Nuzhat (1998) pointed out that noise pollution in Peshawar is also above the maximum permissible limits. The silence zones are not protected from this kind of pollution.

Harabidis, (2008) has pointed out several problems related to noise were found in the school children exposed to noise. Three hundred thirty eight noise exposure samples were collected from 133 construction workers employed in 4 construction trades in USA. The study by Neitzel (1999) proved that the construction workers in several key trades were frequently exposed to noise levels that have been associated with hearing loss. Davies (2009) concluded that chronic exposure to high levels of noise may be associated with increased risk of cardiovascular disease. Use of adequate hearing protectors was also considered important.

Lahore is the second largest city of Pakistan. This is also suffering from the menace of noise pollution but no such studies have been conducted in the city yet. I have chosen this topic to highlight those areas of the city where people are suffering from diseases and other issues caused by noise pollution.

There are a number of factors, which have their influence on noise pollution in the City such as varying traffic conditions in different parts of the city, high concentration of vehicles and the speed at which the traffic is allowed to flow. An important role for the increase in noise pollution of various sources in the city has been carried out by various types of road vehicles. There are certain vehicles which create a lot of noise such as trucks and buses etc.

The population of Lahore district was 6,318,745 in 1998 (District Census Report, 1998). In the mid of 2006 government estimated it approximately 10 million. The number of motor vehicles also increased with the passage of time due to increase in population. This resulted in the increase in levels of noise pollution in the city. People living in the city are also suffering from hearing loss and other noise related health issues.

1.2 Aims and objectives of the study

As mentioned above that no research on noise pollution has been carried out especially in Lahore city. The aim of this research is to study the spatial pattern of noise and see its effects of noise pollution in the city. The study will be carried out by keeping the following objectives in view:

a. The main objective of this research is to see the spatial pattern of noise pollution in the city. There are certain locations in the city where noise level crosses the maximum permissible limits. This research will highlight those locations. In this research GIS techniques will be used to present the patterns of noise pollution.

b. The second objective of the study is to see the effects of noise pollution on human health especially on hearing.

c. The response of people will be gathered through questionnaire, which will highlight the problems of people living in areas of high noise levels.

1.3 Study area

Since 1947, Lahore's growth has resulted in major expansion in south and southwest directions. The city's expansion on the east is hampered by close proximity of international border with India and on the north and west by River Ravi.

Besides the housing development, industrial development has also taken place with the establishment of industrial estates while small scale enterprises has come up in the old city and its immediate environs. The overall cityscape can be described as sprawling low density settlements.

An urban area cannot be planned in isolation. This is particularly true in case of Lahore, which being the provincial capital and a metropolis has a vast area of influence, spreading well beyond the district/provincial boundaries. With the creation of LDA, LMA (Lahore Metropolitan Area) boundary was delineated in May 1975 and was later extended towards south in January 1988 based on the growth trends in the southern corridor. Delineation of LMA boundary was effected for better development control and channelizing urban growth. District Lahore covered most part of LMA, LDA, LCDG, Cantonment Board and DHA. The general altitude of the area is about 208-213 meters above mean sea level. Figure 1.3 shows urban area.

Figure 1.3: Urban area of Lahore City

1.4 Topography

The Lahore Metropolitan Area is generally flat and slopes towards south and south-west at an average gradient of 1:3000. The low lying area is along River Ravi and the comparatively upland area is to the east away from Ravi. The low lands are generally inundated by the river water during monsoon floods. River Ravi flows in the west of Lahore District forming a boundary with Sheikhupura District.

The original physiographic features like remnants of channels and levees have been destroyed or changed by the construction of urban infrastructure. Flood plains have been confined by construction of embankments (bunds), spurs etc. Meandering channels have been replaced by sewerage drains. The sub-recent flood plain is 4-8 meters higher than the recent flood plain and can be identified at number of places i.e. Shalimar Garden, Mughalpura and Multan Road.

1.5 Temperature

Lahore experiences extremes of climate. The summer season starts in April and continues till September. The hottest months are May, June and July. The mean maximum and minimum temperatures during these months vary between 47 degree Celsius and 27.4 degree Celsius.

The winter season lasts from November to March. The coldest months are December, January and February with minimum and temperature reaching up to almost freezing point for a few days. The mean maximum and minimum temperatures for this period are 22 degree Celsius and 5.9 degree Celsius respectively.

Figure 1.4 shows that temperature remains low in the month of January and February. It becomes moderate in the month of April, becomes hot from April to July. Monsoon spells in July and August decreases the mean monthly temperature. A gradual decrease in temperature can clearly be seen in the months from September to December.

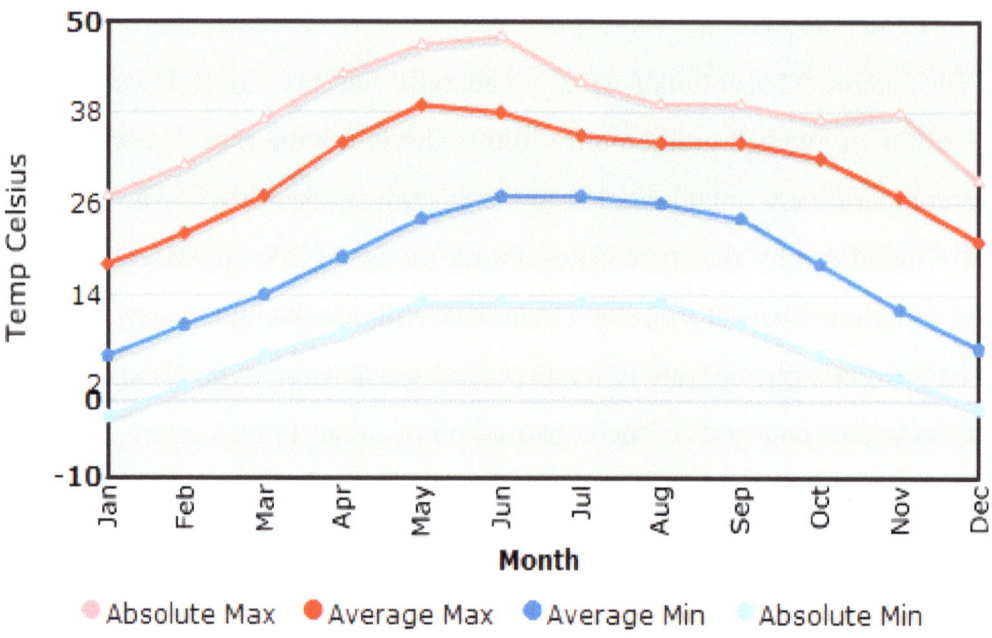

Figure 1.4: Temperature variations in Lahore – The chart plots the average high and low temperature for each month of the year. It also shows the maximum and minimum recorded temperatures.

1.6 Rainfall

Figure 1.5 shows that the amount of precipitation in the month of March and May was only up to 1 mm, which rose to 5mm during the months of June and July.

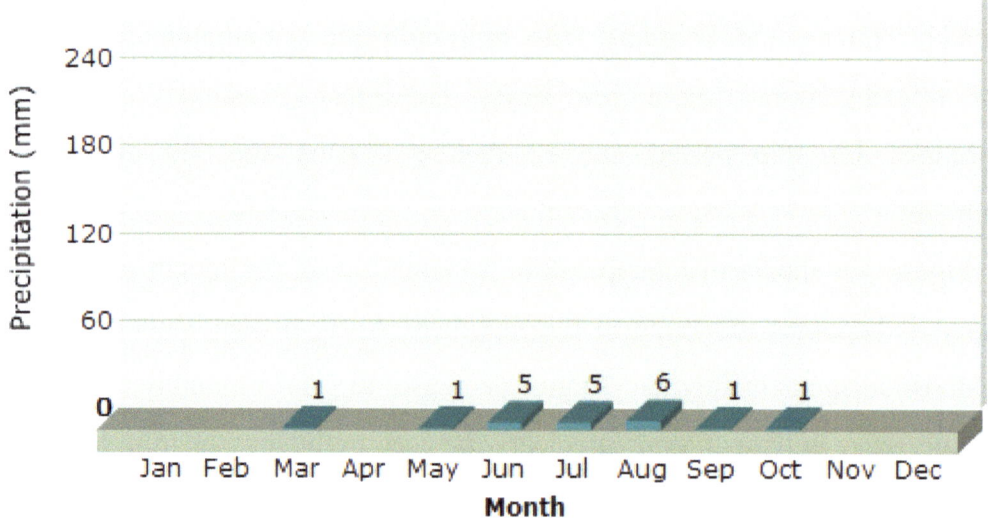

Figure 1.5: Precipitation variations in Lahore – The chart plots the average monthly precipitation amount.

August shows the maximum amount of rainfall, which is 6 mm. The months of September and October show that there will be a clear reduction in the amount of precipitation.

Figure 1.6 shows the average number of rainy days in different months of the year 2011. It shows that there were 5 rainy days of January and February. There were six days in March, five days in April and May, six days in June, ten days in July and August. The expected days of rainfall in September are six, two days in October, one day in November and two days in December.

Towards the end of June or beginning of July the Monsoon season starts which is characterized by heavy downpour and humid weather. It practically becomes oppressive in July, August and September.

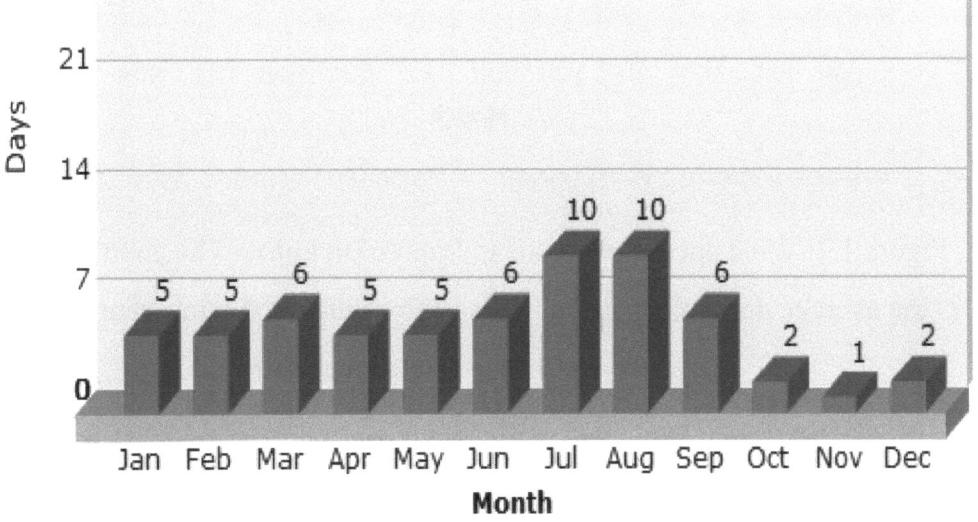

Figure 1.6: Number of rain / drizzle days in month

1.7 Wind speed and direction

On the average 60% of the days during a year are calm when wind movement is negligible. Winter months are mostly calm with minimum wind storms. Wind storms are more common during the months of April to July with maximum occurrence recorded in June when low air pressures are developed due to high temperature.

In winter, the wind mainly blows from north-west while in summer it is from the opposite direction i.e. from south-east, which brings the monsoon rains. Figure 1.7 shows that the wind blows with the highest speed in the month of May. The speed of the wind was 96 kph in May. The lowest wind speed will be in the month of November.

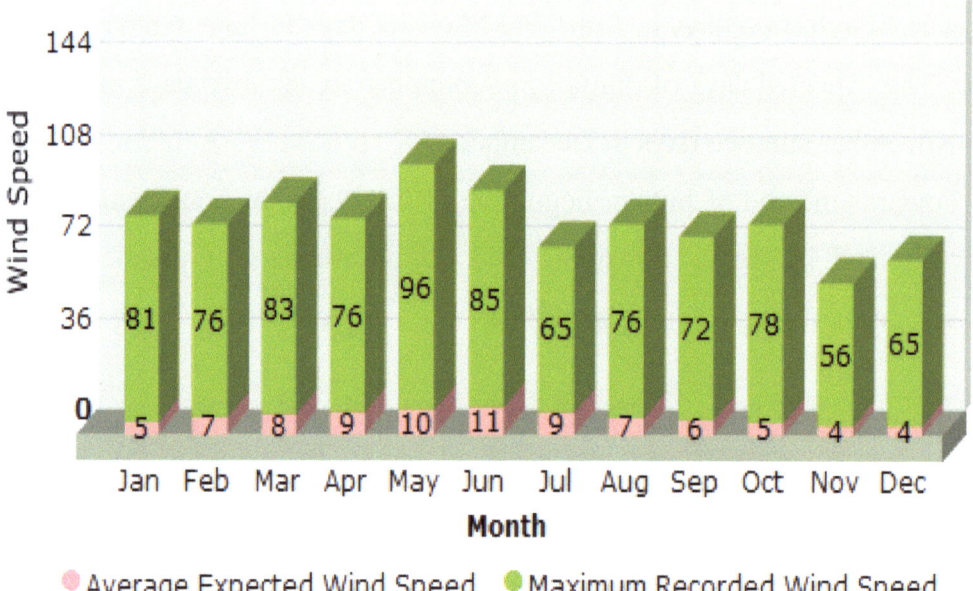

Figure 1.7: Wind speed variations in Lahore (in kph) – The chart plots the average daily wind speed. It also shows the maximum recorded sustained wind speed for each month

1.8 The road network

Major roads connecting Lahore to the other cities are NH5 Multan Road, Raiwind Road, Ferozepur Road, and Jaranwala Road. The main railway line connects Lahore to most of the settlements along northern and southern routes and also to the neighboring country-India, through Wagha in the east.

Primary road network comprising of arterials and collectors is well developed in Lahore as compared to any other historical city. Historically the shape and pattern of arterial network has largely taken place because of geographical location of the city with respect to River Ravi and other major towns in the region including

Peshawar, Rawalpindi and Sialkot in the north, Chiniot, Mainwali and Sheikhupura in the west and Multan in the south west. Collectors have established with the development of Cantonment and the railway station in the east, Secretariat and GOR in the south during British Regime and a number of planned schemes during the post-independence period in the south.

Figure 1.8: Road network of Lahore city

Present shape of the road network developed through historical growth of the city is primarily radial and suits for efficient operation and coverage of public transport but density of primary radials being too low and in the absence of adequate number of distributors/inter-radials, there is unnecessary traffic pressure on the primary network.

Secondary and tertiary road network is generally below any acceptable standard either because of lack of maintenance or at places it is missing due to uncontrolled growth of the city. There has been tremendous effort to improve the road system through politically driven and through well-programmed interventions during last one decade.

1.9 Status of vehicular traffic in Lahore

There were only 39205 vehicles "On Road" in Lahore District in 1974. This number rose to 79382 by 1980, which indicates an increase of over 100% in a period of six years with an average annual increase of 12.5%. This figure rose to 246383 vehicles in 1990, and 561949 vehicles in 1998, indicating and average annual increase of more than 11.4% between 1980 and 1998.

In 2005 the District Officer (Environment) of Lahore had estimated that there are 1.5 million registered motor vehicles in the city. According to Excise and Taxation Department, 900,000 new vehicles were registered in the City between 2002 and 2007 (Alam, 2008).

From the above comparison, the challenge ahead for the city/traffic planners can be very easily assessed, since the road capacities have not been able to keep pace with the growing volume of traffic. Badly managed traffic not only creates traffic congestion, loss of valuable time of passengers and wear and tear of the engine, but also produces increased quantities of Carbon Mono-oxide (CO), which is injurious to human health. Some improvement in the City's main roads was carried out in the recent past especially in posh areas and around Gulberg, but the ever increasing traffic loads demand a multiple approach.

1.10 Impacts of noise pollution on urban life

A major source of urban noise pollution is mass transit as well as other means of transportation. Noises from motor vehicles include horns, alarms, and tire contact with the road and engine acceleration. The strategies to decrease noise pollution can improve the quality of life among urban dwellers.

With the rapid increase in population of the city coupled with the growing living standards, the number of vehicles moving on the roads of Lahore has increased manifold. The vehicular traffic combined with the uncontrolled emissions from the industrial units located in and around the city has created an adverse impact on the quality of the city.

It has been found to interfere with our activities at three levels:

1.10.1 Audiological impacts

At audiological level noise affects the mechanism of hearing and causes hearing loss.

1.10.2 Biological impacts

At biological level noise interferes with the biological functioning of the body such as it can cause headache.

1.10.3 Behavioral impacts

At behavioral level, noise causes depression and annoyance hence it affects the sociological behavior of the people.

Chapter 2:
Literature review

2.1 Literature review

A number of studies over the years have been conducted to investigate the increasing levels of noise pollution at various locations within Pakistan and around the World using varieties of instruments. A review of these previous studies provided insight into the research design used for the current study. The previous studies are briefly discussed in this section to provide background and context for the current study.

Stephenson (1968) pointed out that Traffic was the main source of noise in Central London. He gave details in two experiments. In the first investigation the noise levels were measured due to 1100 vehicles individually under similar conditions and in the second case traffic noise was measured at 140 sites note being taken of traffic volume and composition. The importance of lorries and busses in contributing to high noise levels is discussed as are effects of speed. Urban motorway will have major influence on the noise environment of the future and measurements near existing motorways will have are reported both with respect to traffic volume and to distance from motorway.

Mehdi (2010) confirmed that high road density areas such as the old city of Karachi was characterized with very high level of noise exposure at most times. The average values of traffic noise in mornings, afternoons and evenings exceeded 66dB that could make people feel seriously annoyed according to the WHO outdoor environmental noise exposure guidelines. The peak level noise of traffic was very serious in Karachi. Field measurements showed that the maximum peak noise values exceeded 100dB at the selected sites, which is close to the level of 110 dB at which hearing impediments may occur. Chronic exposure to the peak noise levels can adversely affect the Karachi residents.

Goswami (2009) described that the transportation sector is one of the major contributors to noise in urban areas. The traffic noise environment in Balasore, a city of Orissa, India in terms of standard noise indices, community response and community health effects are worked out. Noise pollution is assessed in six different squares of the town. It is inferred that the noise levels are more than permissible limit in all six investigated locations of the city. A preliminary survey adopting questionnaire method amongst 212 local inhabitants also carried out to gather secondary information about the suffering of noise related health problems.

In a temporal and spatial distribution of road traffic induced noise pollution in an urban environment by monitoring and mapping at Asansol city of West Bengal, India, a total of 35 locations were selected for collection of data, classified as industrial, commercial, residential, sensitive and mixed areas according to the national regulatory standards. Noise recordings were conducted during morning and night hours. Day time Leq level ranged between 51.2 and 89.0 dB (A), whereas it ranged between 43.5 and 81.9 dB (A) during night. The traffic noise index was 80.62 ± 15.88 dB (A) (Range: 49.4-115.8). The computed data were mapped by utilization of Geographic information system methodology that allowed the visualization and identification of the extent and distribution of sound pollution across the study area. This proved to be an ideal tool for carrying out noise impact assessments in urban settings. The study revealed that present noise level in all the locations exceeded the prescribed limit. Based on the finding, it was mentioned that the population in the industrial town was exposed to significantly high noise level, which was caused mostly due to road traffic. The study revealed that vulnerable establishments like schools and hospitals were subjected to significantly high noise level throughout the day (Banerjee D, 2009).

Serkan Ozer (2009) summarized that the noise pollution is the candidate for most important environmental problem of Turkey. The outcome of the study showed that noise values reached up to significant levels even in a small sized City. City centers are the places where noise problems are very high. Similarly noise level in Tokat City was also found to be high, which increased by a ring road passing through the city centre. In addition, long and large vehicles should not pass through the city centre. Routes of public transportation vehicles should be recognized so as not to be accumulated at one point in the city.

Rajesh (2006) confirmed that the noise levels in Dhaka City remains 75-80 dB all over the day and among the 13 problematic persons 3 have NIHL, 3 persons have presbycusis and remaining 7 persons showed unknown reasons, which means 10.34% out of 29 samples had noise induced hearing loss.

Al-Ghonamy (2010) evaluated and analyzed the road traffic noise in the city of Al-Dammam All measurements were taken on an A weighting frequency networks, at a height of about 1.5 m from the ground and on the fast range time weighting. The A weighting characteristics and fast range were simulated at human ear listening response. Day-night and day time noise maps were drawn with the districts being classified into moderately high and extremely-high risk zones in accordance with noise risk levels. Mean day-night range from 68.1 to 90.6 dB (A) and exceeded the permissible environmental standards used in Kingdom of Saudi Arabia

The spatio-temporal distribution of noise was studied in West Bengal generated by urban traffic, by means of monitoring and mapping as a tool for evaluation of impacts (Banerjee 2009).

Lallan singh (2010) confirmed with quantified data that the city is exposed to noise levels ranging mostly from the moderate to extremely high levels in comparison to the national standards. The study advised that immediate administrative and technological mitigative measures should be adopted immediately to prevent auditory and non-auditory health impacts on the local population. This study also proposed certain control methodologies such as control of noise at source of generation and control in the transmission path by installation of barriers between noise source and receiver. Noise pollution is undoubtedly an outcome of urban industrial technology revolution, which has resulted in complex man-environment relationships associated with large scale and speedy exploitation of resources.

Craik (1986) pointed out that many people consider an intense sound to be noise but defining physically noise does not just constitute only intense sound rather it is also a complex sound with little or no periodicity, the sound can be additionally intermittent, that of multi-frequency and impulsive in nature.

Smith (1996) summarized that the sound is an aural sensation caused by pressure variations in the air which is always produced by some sources of vibration. They may be from a solid object or from turbulence in a liquid or gas. These pressure fluctuations take place very slowly, such as those caused by atmospheric changes or very rapidly and be in the ultrasonic frequency range. The velocity of the sound is independent of the rate at which these pressure changes take place and depends solely on the properties of the air in which the sound is travelling. Sound is a disturbance which propagates through a medium having the properties of inertia and elasticity. The medium by which audible sound is transmitted is air.

The nature and level of outdoor traffic noise in an actual urban situation and to verify the relationships between level of traffic noise, traffic volume and traffic composition. Noise measurements were performed at 70 locations uniformly distributed over the town, in the autumn of 1974. A ten-minute record was made at each site ever), hour for 23 hours. The results are presented and compared with published data from previous surveys carried out in other European and North American towns (Benedetto, 1977).

Harman (1973) conducted a noise survey within the Portsmouth City boundaries. Measurements were made throughout the 18-hour day at 33 sites which covered a wide range of traffic conditions. Comparisons were made between the published noise prediction methods and the measured results for sites adjacent to roads carrying free-flowing traffic.

Florentina (2008) asserted that GIS is a tool to preset noise data. The modified Nordic Prediction Methods and analysis of the result was implemented in ArcMap

9.2. a model for road traffic noise was developed in which GIS, which was used as a tool of noise data presentation (Mehdi., 2002).

Harabidid (2008) pointed out high risk zones in a study. The effects of noise pollution on humans have been intensely carried out. A comparison was made between the children exposed to noise to those not exposed to noise.

Harris (1979) reviewed in a study that the blood pressure of children was found to rise by 4-8 mmHG and their ability to learn and to discriminate words was also affected. The performances of younger children deteriorate more as compared to older subjects. Motor and psychomotor functions may get affected under condition of noise exposure.

Netherland government (1987) has taken the following measures to reduce hearing loss due to occupational noise. Noise levels higher than 80 dBA must be considered hazardous and hearing protectors should necessarily be provided by the employer. For noise levels exceeding 85 dBA, technical and organizational measures have to be taken to reduce noise. For noise levels higher than 90 dBA, hearing protectors must be used.

Lords (1963) pointed out that introvert people seems to be more bothered by noise than extroverted individual prone to depression, hypertension and anxiety or who face emotional experience such as divorce or unemployment tend to be more sensitive to extraneous sounds and consider them noise. The ability to hear means being capable of detecting sounds within the frequency range of 16-20,000 Hertz. Subjected to 45 dB of noise, the average person cannot sleep. At 120 dB ears register pain, but hearing damage begins at a much lower level about 85 dB. The duration of exposure is also important. There is evidence that among young Americans hearing sensitivity is decreasing the year by year. Apart from hearing loss, such noise can cause lack of sleep, irritability, heartburn, indigestion, ulcer, high blood pressure and possibly heart diseases. One burst of noise from a passing truck is known to alter endocrine, neurological and cardiovascular function in many individuals prolonged or frequent exposure to such noise tends to make the physiological disturbance chronic.

In addition noise induced stress creates severe tension in daily life and contributes to mental illness.

The effects of noise into auditory effects such as noise induced hearing loss that may be temporary or permanent and non-auditory effects such as increased heartbeat, indigestion, tension, anxiety, emotional imbalance and fatigue. Threshold shift may be temporary or permanent (Tripathy). The noise problems and complaints increased dramatically by the end of 19th and the beginning of 20th centuries as U.S and European societies became more urbanized and mechanized. With time the problem of noise was taken up both socially, politically and legislative and control measures were introduced to reduce noise pollution (Hay, 1982).

There are numerous clinical symptoms and signs like changes in blood pressure, heartbeat etc. have also been attributed many a times to noise exposure. Nausea, headache, insomnia and loss of appetite are also said to be caused by excessive noise exposure. Continuous exposure of noise has been found to cause constriction of blood vessels in humans which may eventually lead to heart ailments (Lehman, 1956).

Road traffic noise in urban areas reduces the quality of residential environment, which results in reduction of property values. This imposes a major cost on society through the need for sound-insulation of buildings. The damage caused by noise pollution can also include productivity losses, health care cost loss of psychological wellbeing. Productivity losses may be caused by exposed person inability to concentrate, by communication difficulties at work or due to fatigue due to lack of sleep (WHO, 1995).

EPA (2006) summarized that in comparison to other pollutants the control of environmental noise has been hampered by insufficient knowledge of its effects on human and lack of defined criteria. In our cities due to poor planning, road construction and progress work on the development of utility infrastructures along the road; the problem of noise pollution is being aggravated day by day. Some areas of Islamabad are getting noisier due to high traffic density and lack of traffic management.

Agarwal (2011) pointed out that the impact of noise on community residing near roadside is very prominent. The degree of annoyance was assessed by means of a questionnaire. It was found that among noise generating sources, road traffic was the major source of noise. A health survey reported about 52% of population was suffering by frequent irritation, 46% respondents felt hypertension, and 48.6% observed loss of sleep due to noise pollution. Common noise descriptors were also recorded at all the selected sites. It was found that the Leq values were higher as compared to the permissible values prescribed by Central Pollution Control Board, New Delhi.

A survey was carried out by Thiery (1988) in a car body workshop involving 234 workers divided into three groups according to age and noise exposure duration. Their hearing levels, determined by using a standardized audiometric testing procedure, were compared to those of a reference population not exposed to noise, to those of a population exposed to quasisteady noises at 95 dB(A), and also to ISO 1999-1987.

Basel (1989) summarized that noise is a major and growing form of pollution. It can interfere with communication, increase stress and annoyance, cause anger at the intrusion of privacy, and disturb sleep, leading to lack of concentration, irritability, and reduced efficiency. It can contribute to stress-related health problems such as high blood pressure. Exposure to high noise levels for long time can cause deafness or partial hearing loss.

Evans (1997) asserted that noise interferes with some human activities and if sufficiently intense (above 140 db) can permanently damage the ear. Lower-level continuous intense noises about 90 to 110 dB (A) can cause temporary hearing loss, from which a person recovers after a period of rest in a quiet environment. The amount of hearing loss depends upon frequency and sound pressure level of noise, bandwidth of noise, duration of exposure each day, and number of years of exposures. It is estimated that the number of people in Europe with hearing difficulties is more than the population of France.

A social survey was done collect responses from residents of 27 different sites in the Greater Manchester area. The sites were exposed to noise emanating from (a) freely flowing traffic on urban roads, or (b) motorway traffic, or (c) congested or disturbed traffic flow on urban roads. Existing noise indices were tested on this general sample of traffic flow situations to determine their efficacy in the prediction of community dissatisfaction to traffic noise. No existing index could handle adequately all the traffic flow conditions. When the indices were combined with measures of traffic volume flow between midnight and 6 a.m. a marked improvement in their predictive capability was noted. In particular, extended indices based on L10 (18 hour) and Leq appeared to be useful predictors of community response to all of the traffic flow situations studied in this project (Yeowart, 1977).

Nuzhat (1998) pointed out that noise pollution in Peshawar is also above the maximum permissible level. Road traffic noise was measured at 18 busy locations in 1995, 1996, 1997, and 1998 using Bruel and Kjaer Integrating sound level meter. It was found that road traffic load in Peshawar has increased to a greater extent and is producing noise above the National Environmental Standard for motor vehicle exhaust and noise, i.e., above 85 dB (A). It has shown rising trends during the past four years. The annual rise was not found to be statistically significant. However a significant difference was observed when road traffic noise levels of the year 1998 were compared with those of 1995. Moreover the difference between traffic noise levels in the field as compared to that in the silent zones was found to be statistically significant.

A cross section survey of the population in Delhi State by Kamla (2004) pointed out that main sources of noise pollution were loudspeakers and automobiles. However, female population was affected by religious noise a little more than male population. Major effects of noise pollution included interference with communication, sleeplessness, and reduced efficiency. The extreme effects e.g. deafness and mental breakdown neither was ruled out. Generally, a request to reduce or stop the noise was made out by the aggrieved party. However, complaints to the administration and police had also been accepted as a way of solving this menace.

Public education appeared to be the best method as suggested by the respondents. However, government and NGOs (Non Governmental Organizations) can play a significant role in this process.

This noise pollution level is associated with blood pressure (systolic and diastolic) and heart pulse rate in schools' children. The test sample schools consisted of six different schools chosen randomly in Jenin city. The measured sound pressure levels (SPL) in all tested schools were found to be above the standard international acceptable levels. Strong positive correlation (person correlation coefficient) was found between sound pressure levels in the sample schools from one side and blood pressures (R=0.96 for systolic and R= 0.98 diastolic) and heart pulse rate (R=0.991) from the other side. The average change rate of systolic and diastolic blood pressures were found to be about 4.60 mm-Hg and 2.74 mm- Hg for every 76.86 dB/hr change in SPL values, respectively. Also, the average rate of change of heart pulse rate was found to be about 5 beats/min, which reflected the strong correlation between changes of systolic blood pressure and heart pulse rate (Roba Mohammed Anis Saeed, 2010).

The noise pollution posed major problem in the Varanasi City and left its adverse effects on the exposed people. Noise levels have reached an alarming condition there. 85% of the people were disturbed by traffic noise; about 90% of the people reported that traffic noise is the main cause of headache, high blood pressure problems, dizziness and fatigue. People having higher education and income level are much aware of health impact of traffic noise. Marital status was found to be significantly affecting the annoyance level caused by traffic noise. Traffic noise was found to be interfering daily activities such as resting, reading and communication. Of more than 28 million Americans with hearing impairment, about 10 million were have hearing loss caused by excessive noise exposure. Some noise pollution sources are obvious, but many hazards aren't. And once you lose your hearing, or some portion of it, it never comes back (Downey, 2003).

Among eight noise measurement locations in the twin cities of Islamabad and Rawalpindi, it is amazing to note that at all locations the daily maximum and daily equivalents are higher than the maximum permissible limit of 85 dB (A) of National Environmental Quality Standard (NEQS) for motor vehicle noise at 7.5 meters from the source. Hundred percent of noise measurement locations are exposed to higher noise levels. Even though the daily minimum noise level at all locations is more than 60 dB (A), which clearly reflect an aggravated problem of noise on the road sides of the major cities of Pakistan. The highest maximum noise level 98 dB (A) was found near to Pirwadahi General Bus Stand from where the intra-city heavy traffic operates through the day. Here at this location, the daily equivalent noise level was calculated 97.1 dB (A) from the daily recorded maximum and minimum noise levels. The second highest noisy location among the measured locations was near Choare Chowk, Peshawar road Rawalpindi. Here the daily maximum noise level was 97 dB (A). One of the obvious reasons of high noise level at this location is that the Choare Chowk is the busiest intersection of Peshawar road (EPA Report, 2005).

Kurakula (2007) summarized that the road traffic noise in urban areas reduces the Quality of the residential environment, which results in reduction of property values. The damage caused by noise pollution can also include productivity losses, health care cost and loss of physiological wellbeing. Productivity losses may be caused by exposed person inability to concentrate, by communication difficulties at work or due to fatigue due to lack of sleep.

Stephen (2003) described that noise is a prominent feature of the environment including noise from transport, industry and neighbors. Exposure to transport noise disturbs sleep in the laboratory, but not generally in field studies where adaptation occurs. Noise interferes in complex task performance, modifies social behavior and causes annoyance. Studies of occupational and environmental noise exposure suggest an association with hypertension.

Environmental noise consists of all the unwanted sounds in our communities except that which originates in the workplace. Environmental noise pollution, a form of air pollution, is a threat to health and well-being. It is more severe and widespread than ever before, and it will continue to increase in magnitude and severity because of population growth, urbanization, and the associated growth in the use of increasingly powerful, varied, and highly mobile sources of noise. It will also continue to grow because of sustained growth in highway, rail, and air traffic, which remain major sources of environmental noise. The potential health effects of noise pollution are numerous, pervasive, persistent, and medically and socially significant. Noise produces direct and cumulative adverse effects that impair health and that degrade residential, social, working, and learning environments with corresponding real (economic) and intangible (well-being) losses. It interferes with sleep, concentration, communication, and recreation (Lisa, 2007).

Noise refers to a powerful, unorganized, irregular, or unwanted sound. It is an unpleasant and hazardous sound that may affect work performance and cause physiological and psychological stress. It has a tremendous impact on physiological responses, including tachycardia, hypertension, anorexia, insomnia, thyroxin, and adrenaline increases; delayed wound healing; increased complications; and prolonged hospitalization. Noise may also pose psychological stress and symptoms such as annoyance, impatience, rage, discontent, excitement, frustration, and uneasiness (Suh, 2010).

Dalton (2001) described that the participants were examined in a population-based study of age-related hearing loss in Beaver Dam, Wisconsin. Hearing thresholds were determined by audiometry. Hearing loss was defined as the pure-tone average of the frequencies 500, 1,000, 2,000, and 4,000 Hz greater than 25 dB HL in either ear. Information regarding exposure to leisure-time noise was obtained by interview. After adjusting for potential confounders, individuals who engaged in leisure activities with average sound levels greater than 90 dB (A) were significantly more likely to have a hearing loss than participants who did not engage in noisy leisure activities. Individuals who engaged in woodworking were 30 per cent more likely to have a hearing loss than those who had never done woodworking. There was a 6 per cent increased risk of hearing loss for each 5-year period of participation.

Chapter 3:
Noise pollution and its effects

3.1 Introduction

The term noise has been derived from a Latin word nausea meaning sea sickness (Singal, 2005), implying unwanted sound (Encyclopedia Americana) or sound that is loud, unpleasant or unexpected (Katyal, 1989), get dumped into the atmosphere without regarding to the adverse effects it may be having (Chhatwal, 1989). Psychologically, noise is just any sound undesirable by the recipient and may adversely affect the health and well-being of individuals or populations. It is wrong sound, in a wrong place, at the wrong time (Singal, 2005).

The extent to which noise contribute to the deterioration of our environment cannot easily be determined as that of pollution from other sources. Different people are not equally affected by noise pollution. There are different factors which have direct influence on the listener while considering a sound to constitute noise such as time, place and mood of the listener is also important. At one time and place the sound produced by one instrument may be noise for an individual but may be a rhythmic sound for another. If an individual feels happy he may not think a sound of a high intensity as noise but if one feels sad, may take that much intensity of sound as noise (Kupchella, 1993).

If we treat sound as an auditory sensation produced by these vibrating bodies, it gets characterized by its pitch, loudness and tone quality. These characteristics cannot be measured because they have been psychological sensations and depend upon the ear and judgment of the individual observer. In physical terms, sound may be defined as a fluctuation in pressure in an elastic medium, which give rise to the objective characteristics of sound such as frequency, intensity and wave form.

3.2 Measurement of noise

The simplest method of noise measurement consists of assessing linear sound pressure level (SPL) at any time, disregarding variations with time, over a broad frequency band covering the whole of the audible frequency range (Singal, 2005).

3.3 Unit and range of noise

Sound is measured by several complex systems, but the best known unit of measurement is the decibel (dB), a unit named after Sir Alfred Bell. The decibel is a tenth of the largest unit, the bel.

One decibel is equivalent to the faintest sound that can be heard by human ear. Frequency of the sound is defined as the number of vibrations per second. People can hear sound from 16-20,000 Hertz, but this range is reduced with age and other subjective factors (Katyal, 1989).

The range of vibrations below 16 Hz are infra-audible and those above 20,000 Hz are ultrasonic (Kumar, 1999). Some persons can hear frequencies that others are not able to detect. Many animals can hear sounds inaudible to human ear.

Human ear is known to be sensitive to extremely wide range intensity from 0 to 180 dB. While 0 dB is the threshold of hearing and 140 is the threshold of pain. Some people feel discomfort even with sound of 85 dB. Sound pressure level is a measure of air vibration that makes up sound.

Because the audible sound range by human ear is wide $2*10-5-200$ Pascal or 100 dB and above where threshold of pain starts. Table 3.1 lists sound levels in decibels.

Table 3.1: Sound pressure levels for various sources

Sources	Sound pressure level in dB	Remarks
Rustling of leaves	0	Threshold of hearing
Broadcast studios	20	
Bedroom in home and library	40	
Residential area without traffic	40	
Refrigerator	46-68	
Washing Machine	48-78	
Window A/C unit	60-73	
Vacuum cleaner	60-85	
Food blender	62-88	
Automobile	60-90	
Train	72-91	
Heavy Truck	75-89	
Inside concert hall during performance	80	
Musical Band Party	84-93	
Motor cycle	80-100	
Silencer less Scooter	92	
5 meters away from car horn	100	
Construction noise	110	
Hydraulic press	130	Threshold of Pain
Jet takes off	150	
Large rocket engine	160-180	

Source: (Kumar and Katyal, 1999 and 1989).

Sound pressure levels are expressed in units of pressure (force per unit area) and are defined mathematically as:

$$SPL = 20\log_{10}(P/P_0)$$

Where

P is the measured sound pressure.

P_0 is the reference pressure which is equivalent to 2×10^{-5} Pa.

The factor 20 appears in the equation due to the fact that energy or intensity of the sound wave is proportional to the square of the amplitude of sound waves (WHO,

1999). In the field of noise pollution (especially when noise pressures become noisy) several physical quantities and notations are being used:

(a) LNP: Noise pollution level in dB, also written as NPL
(b) L_{eq}: Equivalent continuous Sound Level in dB. It is energy equivalent sound level, represents the statistical average for any fluctuating sound occurring during a particular time period.
(c) L_{10}: The noise level in dB exceeded 10 % of the measured time.
(d) L_{90}: The noise level in dB exceeded 90 % of the measured time.

These quantities can be related to each other as below:
$$LNP = L_{eq} + L_{10} - L_{90}$$

3.4 Sources of noise pollution

There are three sources of sound by category. These are Point Source, Linear Source and Plane Source.

3.4.1 Point source

A sound source can be considered as a point source, if its dimensions are small in relation to the distance to the receiver and it radiates an equal amount of energy in all directions. Such point sources are industrial plants, aircrafts and individual road vehicles. The sound pressure level deceases 6 dB whenever the distance to a point source is doubled.

3.4.2 Line source

A line source may be continuous radiation, such as from a pipe carrying a turbulent fluid, or may be composed of a large number of point sources so closely spaced that their emission may be considered as emanating from a line connecting them. The sound pressure level decreases 3dB, whenever the distance to a line source is doubled.

3.4.3 Plane source

A plane source can be described as follows. If a piston source is constrained by hard walls to radiate all its power into an elemental tube to produce a plane wave, the tube will contain a quantity of energy numerically equal to the power output of the source. In the ideal situation there will be no attenuation along the tube. Plane sources are very rare and only found in duct systems.

There are numerous sources of noise in the city but they may broadly be classified into two classes viz industrial noise and non-industrial noise.

3.4.4 Industrial source

The industrial sources may include various industries operating in cities that are associated with manufacturing of machinery, shoes, food, spare parts of cars, motor cycles and trucks.

3.4.5 Non-industrial source

The present study comprises only on the caused by road traffic and in traffic noise 70% of noise is contributed by vehicle noise. Among the non-industrial sources, important are as follows:

(a) Automobiles (road traffic)
(b) Trains
(c) Air Craft
(d) Construction work
(e) Radio, microphones etc

3.5 Mechanism of Hearing

Our ear has an external broad lobed part called external ear or auricle or pinna, which converges the sound waves to an inner tube called meatus or auditory canal. The tube ends at tympanic membrane or ear drum which is very thin and tough of about 1 cm diameter. Sound waves create vibration in the membrane and the vibratory motion is transmitted inside by three small ossicles to the cochlea. The mechanical impulse reaching the cochlea are converted into electrical impulses though numerous cilia and get transmitted through auditory nerve to brain. The ear is connected to nasal passage through eustachian tube to adjust the atmospheric pressure to a steady force on the ear drum. There is a semicircular canal which controls the equilibrium or balancing mechanism. Thus high pressure sound can cause damage in a number of ways on our hearing ability, brain and balancing mechanism.

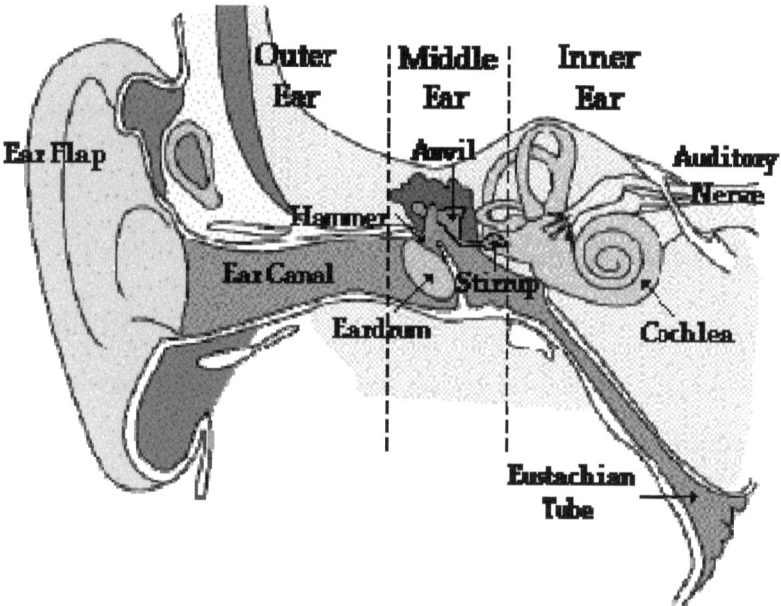

Figure: 3.1. Parts of human ear.

As a result of evolutionary process our ear has built in mechanisms to perceive from very feeble sound to hear the powerful thunder. With aging there is a gradual loss in hearing ability.

Persons working in noisy places lose it faster. Thus the hearing ability of an average village dweller is much better than the corresponding age person living in a noisy city or working in a factory. Of course to a certain extent the ear apparatus is capable of reducing high pressure noise impact on the ear drum, but beyond that upper threshold level, the noise causes pain. Damage to ear drum and ossicles causes deafness.

To some extent it is not permanent but loss of cilia in the Corti, which are responsible for conversion of sound waves into electrical impulses, causes permanent loss of hearing.

3.6 Effects of noise

There are many problems, which are caused by noise pollution such as the sound in range of 120-150 decibels can affect the respiratory system and affect balance to the extent of dizziness, disorientation, nausea and vomiting.

Sound may produce a number of reflex-like reactions. The most common of these is the startle response which is usually caused by loud, unexpected sounds and is evident as a flexing of various muscles and a blinking of the eyes. These reactions in themselves do not have any known effect on health, although in many work situations they could startle a person into injuring himself or others. Physiological changes accompanying other noise exposures include a vascular response characterized by peripheral vasoconstriction, changes in heart rate and blood pressure, various glandular changes (such as increased output of adrenalin) evidenced as chemical changes in blood and urine a change in the electrical resistance of the skin with changes in activity of the sweat glands and brief changes in skeletal muscle tension.

Another major area of concern is the effect that noise may have on the performance. It is likely that any new sound or change in an existing sound may result in at least momentary distraction, and that this may impair a person's ability to perform some tasks. More generally, changes via the non-specific connections in the cortex associated with prolonged exposure to even relatively steady noise may have some effect on such task performance.

Another direct effect of noise is on hearing. The problem of noise induced hearing impairment begins to occur somewhat between 80 and 90 decibels (Kupchella, 1986). Normal hearing is the ability to detect sounds in the audio frequency range. However, it is a common thing that hearing diminishes with age (a condition known as presbycusis).Hearing loss or impairment is the measure of the reduction in the ability to hear from that of a normal person. It tends to be small up to the age of 30 years or so but increases rapidly with increasing age. Low frequencies below 1000 Hertz are relatively little affected but loss increases rapidly with frequency. In addition to presbycusis every day exposure to noise also causes hearing loss known as Noise Induces Hearing Loss (NIHL).

Noise Induced Temporary Threshold Shift (NITTS) can vary in magnitude from a few decibels restricted to a narrow region of frequencies to the extent of temporary deafness. Noise between energy concentrations between 2000 Hz to 6000Hz produce greater temporary threshold shift. In general noise exposure of A-weighted sound level must exceed 80dBA SPL for more than 8 hours a day before a person may experience temporary threshold shift (Singal, 2005). To protect the majority of people from being moderately annoyed during the daytime, the sound pressure level should not exceed 50 dB LAeq. In schools and preschools, to be able to hear and understand spoken messages in class rooms, the sound pressure level should not exceed 35 dB LAeq during teaching session (Berglund. B, 1995).

Hearing plays an important role not only in speech communication but in the perception of a variety of important sounds. It provides information about our environment, and acts as the first warning of impending danger. It can allow identification of the source of the sound, give information about the location of the source, and in a more subtle manner the degree of reverberation gives a 'feel' for the acoustic environment. Noise can impair or eliminate perception of this information.

Standards for hearing health are very misleading. Sound pressure against the ears is measured in decibels (db) on a scale that is logarithmic. That means each increase of 10 db represents a ten-fold increase in noise intensity. In other words, a small sound increase from 90 to 100 db means 10 times the pressure against those delicate hair cells (Downey, 2003). Aberrant, lesion-induced neuroplastic changes in the auditory pathway are believed to give rise to the phantom sound of tinnitus that is ringing of bells and hearing loss (Kraus, 2011). Experiencing a hearing loss sets the foundation for potentially untold anxiety-producing situations (Carmen, 2002).

Speech reception is the most important and also the most complex use of the auditory system. Noise can either mask speech to make it inaudible, or by masking only some frequencies leave it audible but of reduced intelligibility. The interfering effect of noise on face-to-face speech communication can be assessed in terms of the maximum possible distance between the speaker and the listener start in a given steady background noise level for a particular voice level. For personal conversation where a separation of 2 m is typical it can be seen that normal communications are possible in noise levels up to approximately 60 dB(A). At higher levels the talker will unconsciously raise his voice level to compensate for the high-noise environment, and with this greater effort communications may still be just reliable in noise levels up to approximately 70-75 dB (A).

Noise effects on human beings

Noise nuisance
- Efficiency
- Comfort
- Enjoyment

Noise hazards
- Permanent hearing loss
- Neural-Humoral stress response
- Destruction of artifacts

Efficiency
- Mental stress
- Frustration
- Task interference
- Irritability

Comfort
- Sleep interference
- Communication
- Invasion of privacy
- Damage of artifacts
- Habit of talking loudly

Enjoyment
- Concentration interference
- Meditation interference
- Recreational interference
- Temporary hearing loss

3.7 Environmental Quality Standards (EQS) for noise in Japan

The duration of day time in Japan is considered to be from 6:00 a.m. to 10:00 p.m. The area is divided into different categories.

The category AA is applied to areas where quietness is requires such as hospitals and educational institutions. Area category A is applied to areas used exclusively for residence. Area category B is applied to areas mainly for residences.

Area category C is applied to areas used for commerce and industry. Source: http://www.env.go.jp/en/air/noise/noise.html (date of access, 23-08-2011).

Table 3.2:

Area Category	Standard Values	
	Day Time	Night Time
A	60 dB or less	55 dB or less
B = Facing road with two or more lanes and C = Facing road with one or more lanes	65 dB or less	60 dB or less

3.8 The duration of exposure to noise in USA

In residential areas the noise level should not exceed 45 dB during night time. During the day time the permissible noise exposure in USA is given in Table 3.3: Source:http://www.osha.gov/pls/oshaweb/owadisp.show_document?p_table=standards&p_id=973 (date of access, 23-08-2011).

Table 3.3:

Duration per day, hours	Sound Level dB slow response
8	90
6	92
4	95
3	97
2	100
1	105
1/2	110
1/4 or less	115

3.9 National Environmental Quality Standards (NEQS) for Noise

According to the national standards of Pakistan the noise levels in residential areas should not exceeds 65dB during the day time and 50 dB during the night time. In commercial areas, noise level should not exceed 70 dB during day time and 60 dB during night time. While in the silence zones (100 meters areas around hospitals, educational institutions and courts) it should not exceed 55dB during day time and 45dB during night time. (The gazette of Pakistan, 2010).

Noise pattern in the City does not follow the national standards and shows that it has reached an alarming situation in which if it is left unchecked can cause severe damages to human lives. Many countries have implemented new technologies to control noise pollution in urban areas. For example, low noise generating engines, changing in quality of tyres and changes in road material. These technologies have proven to reduce the noise on individual scale. However, the overall noise pollution in urban areas is still increasing. The duration of exposure of people in the City to high noise levels is 12 hours.

The extent to which noise contributes to the deterioration of our environment could not be as early determined as that of pollution from other sources because different people are not equally affected by the same noises. There also occur a vast variation in individual's sensitivity to noise and people are affected differently when they are at home and when they are outside or at work.

Chapter 4:
Data sources and methodology

4.1 Introduction

Traffic noise need to be measured in order to study the diseases associated with it. Hearing loss, loss of appetite, depression and sleeplessness are caused due to noise pollution. This chapter provides a conceptual frame work of the study.

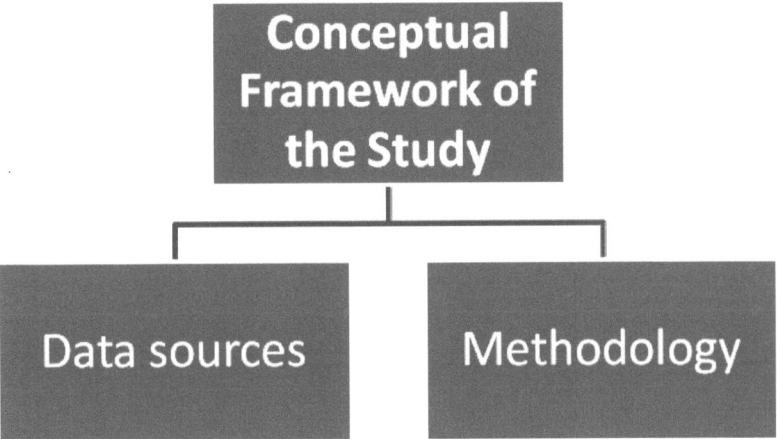

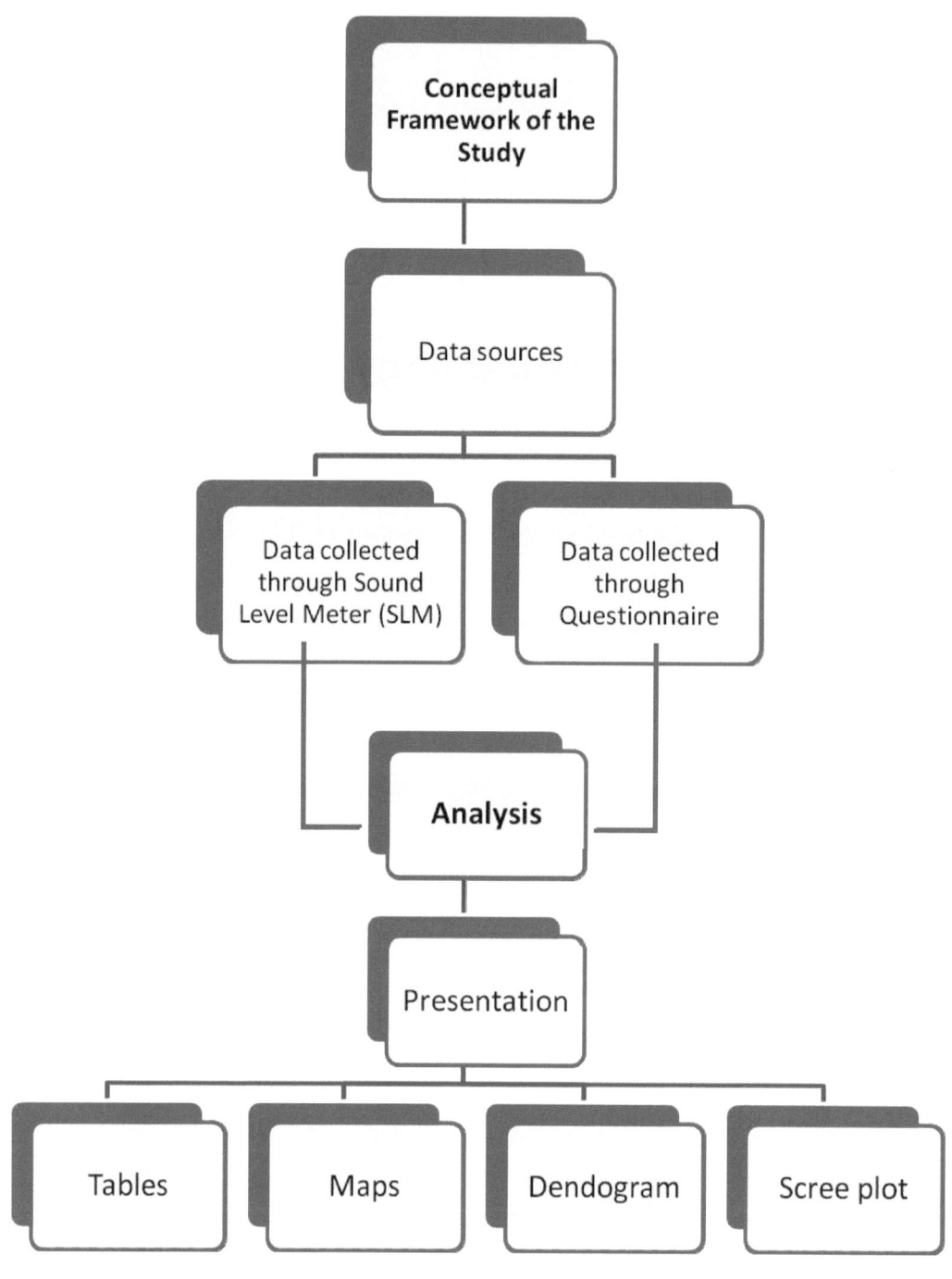

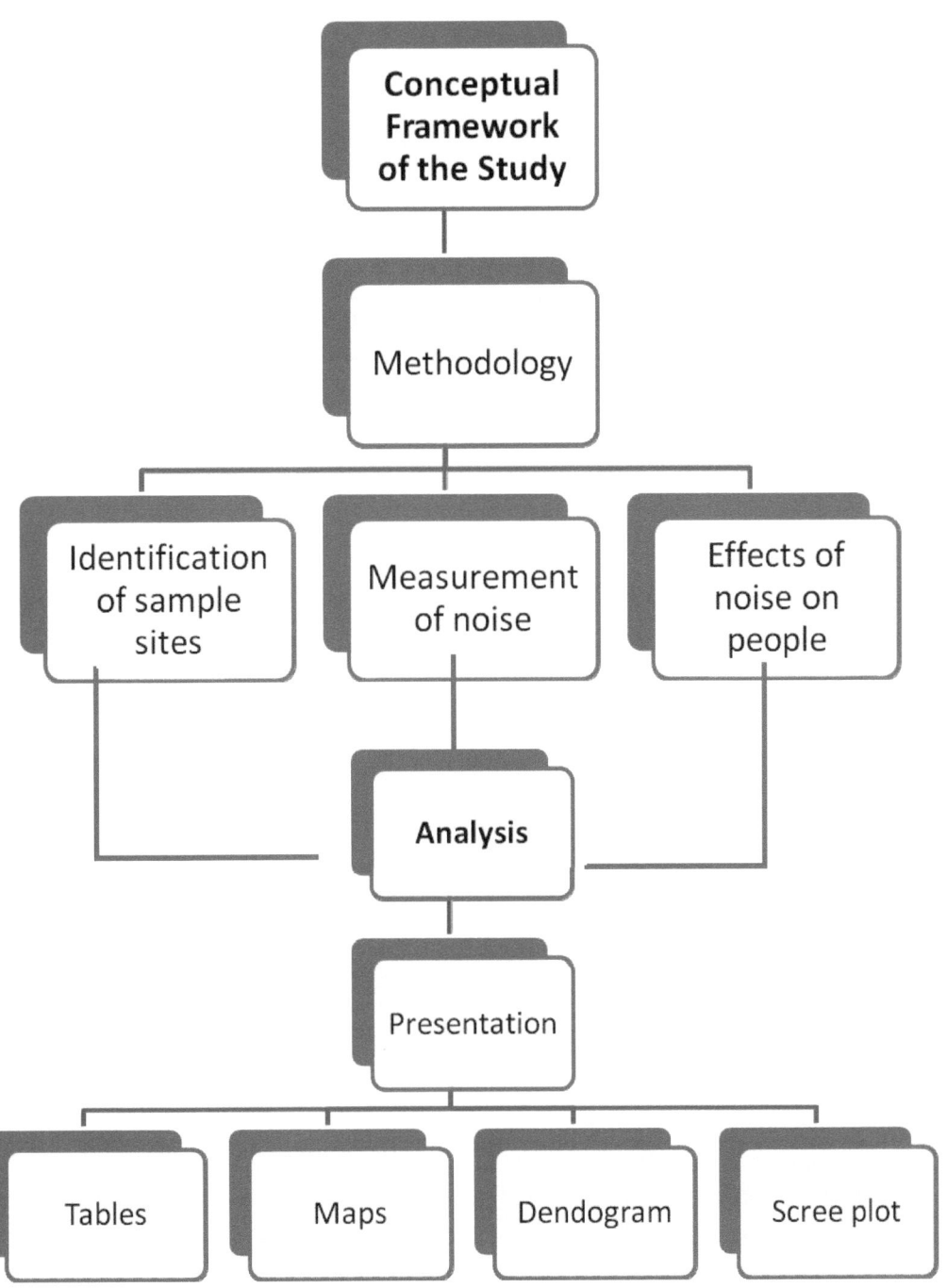

4.2 Data Collection

4.2.1 Data Sources

Data collection is an important work and is the back bone of research. Appropriate data is necessarily required for a proper analysis, which leads to derivation of accurate results. The data base was prepared on the basis of the information collected from different sources.

Primary data were collected by field survey. The tools used for collection of primary data collection were Sound Level Meter (an instrument, which can measure noise range from 35-130dB). The type of sound level meter has been discussed in the second chapter. The data were collected from the sample sites. Data were collected from different sites using dBA. Measurements were kept as slow response and A weighted sound level was preferred for road traffic noise. The instrument read A grade weight values directly and the microphone was kept 1.2 meter above the ground at a distance of 1.5-5meter. The noise level was measured thrice a day. First readings were taken from 6 A.M. to 8 A.M. Second readings were taken from 12 P.M. to 2 P.M. Third readings were taken from 8P.M. to 10 P.M.

The second source used to see response of the people regarding effects of noise was questionnaire method. The questionnaires were filled by the people at those places where the level of noise was very high. A questionnaire was built related to health impacts of noise pollution and awareness about it among people.

4.2.2 Equipment used for noise measurement

The A-weighted decibel scale begins at zero. This represents the faintest sound that can be heard by humans with very good hearing. The loudness of sounds (that is, how loud they seem to humans) varies from person to person.

The noise levels were measured by sound level meter, which is used to measure sound levels. It is so designed that it responds to sounds in approximately the same way as the human ear will respond to any sound, giving an objective, reproducible sound levels. The sound level meter (SLM) in a way, simulates human hearing as closely as it is possible and practicable. The sound signal is first converted to an identical electrical signal by a high quality omni directional microphone.

Table 4.1: Specifications of a sound level meter

Parameter	Specification
Measuring range	35-130dB
Time response	Slow and Fast
Measuring Mode	Instantaneous, Max Hold and L_{eq}
Output	AC and DC
Power supply	With batteries
Display	LCD
Accessories	Wind screen

Wind on the microphone produces a noise which may seriously affect the accuracy of a measurement. The wind noise was reduced by the use of wind screen. These screens are commonly spherical balls or porous and foamy that fit over the microphone and has negligible effect on the frequency response of the microphone.

4.2.3 Measurement Procedure

For traffic noise problem it is useful to know the equivalent continuous sound level Leq. The Sound Level Meter was kept 1.2 m above from the ground. The noise measurements recorded were Leq, Lmax, Lmin. Values of Lmax give the idea about maximum noise levels measured. Unusually high values of Lmax represent the cases of vehicles honking continuously or the vehicles are without proper silencer. Values of L_{min} represent the minimum noise levels measured.

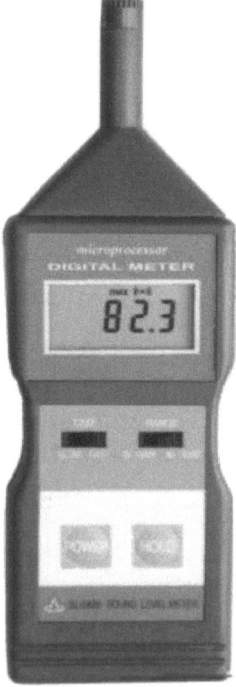

Figure 4.1: Sound Level Meter

4.3 Methodology

4.3.1 Site selection

In order to do this research the first task was site selection. Lahore is the second largest city. Major Roads were taken as primary roads. The roads adjacent to the major roads were taken as secondary roads. The roads adjacent to the secondary roads were considered as tertiary roads. By keeping the large size of Lahore City in view random sampling technique was used to study noise pollution pattern in the area.

4.3.2 Sample location

To cover the maximum area seventy six sample points at each location in the City were selected. These points were selected along the roads.

Table 4.2: Sampling locations

Id	Location	Id	Location
1	A.G.Office (1)	39	Jail Road Canal (39)
2	Ali Town Raiwind Road (2)	40	Jain Mandar (40)
3	Anarkali (3)	41	Jinnah Hospital (41)
4	Assembly hall (4)	42	Kalma Chowk (42)
5	Badami Bagh (Bus Terminal In) (5)	43	Lahore College (43)
6	Badami Bagh (Bus Terminal Out) (6)	44	Lahore Hotel (44)
7	Bhatti Chowk (7)	45	Lane 1 Railway Station (45)
8	Bohr Wala Chowk (8)	46	Lane 2 Railway Station (46)
9	Campus Bridge (9)	47	Liberty Chowk (47)
10	Canal Bank Road (Outside GOR-1) (10)	48	Lohari (48)
11	Chah Blaqi Wala (11)	49	M.A.O. College (49)
12	Chuburji Chowk (12)	50	Mall Road Canal (50)
13	Club Chowk (13)	51	Mayo Garden 1 (51)
14	CMA Colony Cantt (14)	52	Mayo Garden 2 (52)
15	Data Darbar (15)	53	Mayo Garden (Outside) (53)
16	Dharampura Bridge (16)	54	Mazzang Chungi (54)
17	Do-Morya Bridge (17)	55	Minhaj College Township (55)
18	Doctors Hospital (18)	56	Mochi Gate (56)
19	Doctors Hospital (Underpass) (19)	57	Model Town Park (57)
20	Dubai town (By-pass) (20)	58	Muslim League House (58)
21	Eden Villas (21)	59	Neela Gumband (59)
22	Ferozpur Road (Bus Stop) (22)	60	Old Campus (60)
23	Fortress Stadium (Outside) (23)	61	Railway Station (61)
24	Gamey Shah (24)	62	Regal Chowk (62)
25	Garhi Shahu 1 (25)	63	Safan Wala Chowk (63)
26	Gari Shahu 2 (26)	64	Secretariat (64)
27	Garhi Shahu Bridge (27)	65	Shadman Chowk (65)
28	GOR 1 (28)	66	Sha-Alam Chowk (66)
29	GOR 1 (At Distance) (29)	67	Shalimar Garden (67)
30	Government Saleem Model School (30)	68	Sheranwala Gate (Inside) (68)
31	Governor House (31)	69	Sheranwala Gate (Outside) (69)
32	GPO Chowk (32)	70	Simla Hill (70)
33	Gulberg Canal (33)	71	Sir Ganga Ram Hospital (71)
34	Gwal Mandi (34)	72	Taxali (72)
35	Hall Road (35)	73	Thokar Niaz Baig (73)
36	High Court (36)	74	Urdu Bazar (74)
37	Hussain Chowk (37)	75	Yadgar Chowk (75)
38	Imam Bargah (38)	76	Zafar Shaheed Chowk (76)

The maps have been developed by using images as base maps acquired from Google earth software. All the entities e.g. roads, important places, important locations were first digitized in Google earth software and stored in .KML or .KMZ extensions. The KML files then converted into shape files (.SHP) by using ".kml to .shp converter" in Arc Toolbox by using the ArcGIS v9.3 software. The projection system of both files kept same i.e. Geographic Coordinate System to avoid any error/ abruption in the data. Then a layout is designed in Arc Map for the maps to be generated on.

All the surveyed noise pollution points have been collected though Global Positioning System (GPS) by visiting the mentioned locations in the map. The Latitude and Longitude of each single point recorded. A Microsoft Office Excel sheet made to but all the collected data in it and then the same excel sheet joined to the shape file using ArcGIS software's join and relates tool.

4.4 Use of multivariate analysis techniques

4.4.1 Cluster analysis and Factor analysis

Questionnaire method was used to collect information from people regarding causes of noise pollution and the impacts of noise pollution on human health. Factor analysis was done in order to show the results of information gathered through questionnaire.

4.5 Presentation

The information was transferred to GIS (Geographic Information System) database to get its suitable presentation. The data were presented in the form of different maps, graphs scree plots and dendogram.

Chapter 5:
Spatial patterns of noise pollution

5.1 Introduction

Transportation sector is one of the major contributors to noise in the urban areas. The traffic noise environment in terms of standard noise indices, community response and community health effects are worked out in the present study in Lahore city. The investigation of patterns of noise in the city can play an important role in planning and decision making regarding residential colonies and infrastructure.

Noise pollution is assessed at different points of the City. It is inferred that the noise levels are more than permissible limit in all the investigated locations of Lahore city. Lahore is a polluted city as for as noise is concerned. The noise values differ from place to place in the City. The basic noise levels exist for the Lahore city. The use and application of noise mapping by using GIS methodologies has enabled the efficient acquisition, management and elaboration of geo-referenced data and receivers. The major objective of the study was to enumerate the spatial distribution of traffic noise in Lahore city and develop suitable map for evaluation of its impacts.

The traffic noise was measured at seventy six different points along the major roads and within the walled city area as shown in the Figure 5.1. In Asansol city of West Bengal, India 35 locations were taken to study noise pollution (D. Banerjee, 2009). Most of the locations were chosen as sample points by keeping the noise related problems of the people in view.

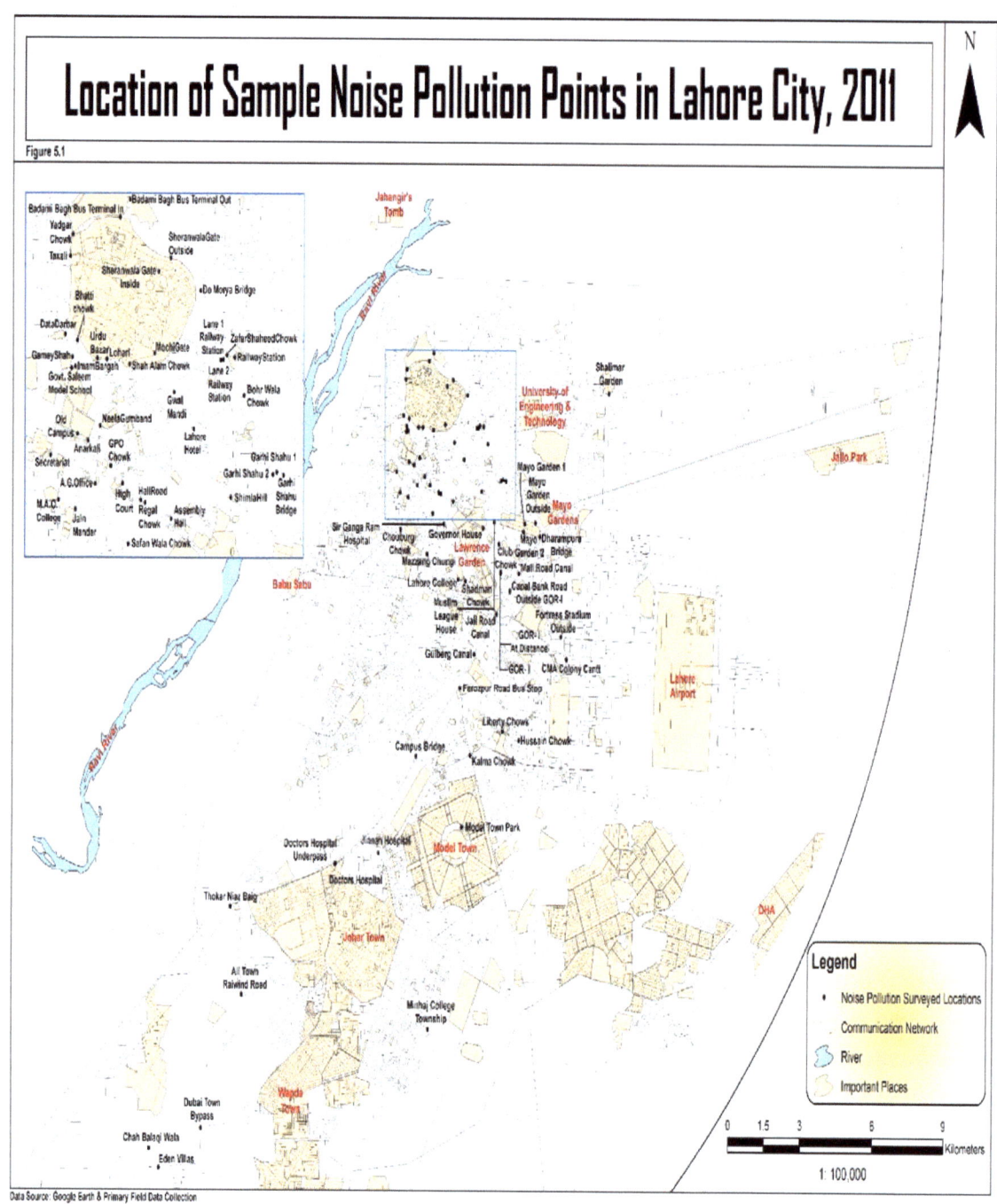

Figure 5.1 Location of Sample Noise Pollution Points in Lahore City, 2011

A few quiet places were also chosen to make a comparison between different parts of the city. All the observations were made thrice a day. The time of measurements was from 6AM to 8AM, 12 PM to 2 PM and 8PM to 10 PM. Averages were taken to show the results. Maximum noise level was observed at Badami Bagh (Bus Terminal Inside), which was 120.4 dB (A). The source is primarily motor vehicular traffic. The permissible level for road traffic noise is 70 dB (A). The minimum noise level was observed at Chah Blaqi Wala, LDA Avenue one and the cause of minimum noise was its remoteness from the main commercial areas. At all locations the noise level has increased due to traffic vehicular noise. The noise measuring instrument used was Sound Level Meter.

Figures 5.2. a and b show the locations of all the sample points. Graduated symbols have been used to show the level of noise pollution. The smallest circle shows those areas where the noise level ranges between 47-70 dB (A). Next higher level of circle represents the areas where the noise level ranges from 71-80 dB (A). The medium sized circle represents areas where noise levels range from 81-90 dB (A). Next higher level of circle represents areas having noise level from 91-100 dB (A). The largest circle shows areas of highest level of noise pollution that is from 101-120. The graph along the map explains the noise values at different points of the city. The map shows that the noise levels are high in the area of walled city, along Jail road, Mall road and Garhi Shahu. The high noise levels at these places create problems.

The different insets on this map highlight the areas of concentration of sample locations. The river has been shown in blue color. All the roads have been shown with black lines. Almost all the sample locations are the rushy places. Whole sale markets and other business activities played important role in increasing noise levels towards walled city, Garhi Shahu and Mall road. Jail road is again another road, which is considered to be a busy road but business activities are different along this road. Many hospitals and colleges are located along this road, which make this road a silence zone but the noise levels violate the limit of noise set by Pakistan Environmental Protection Agency.

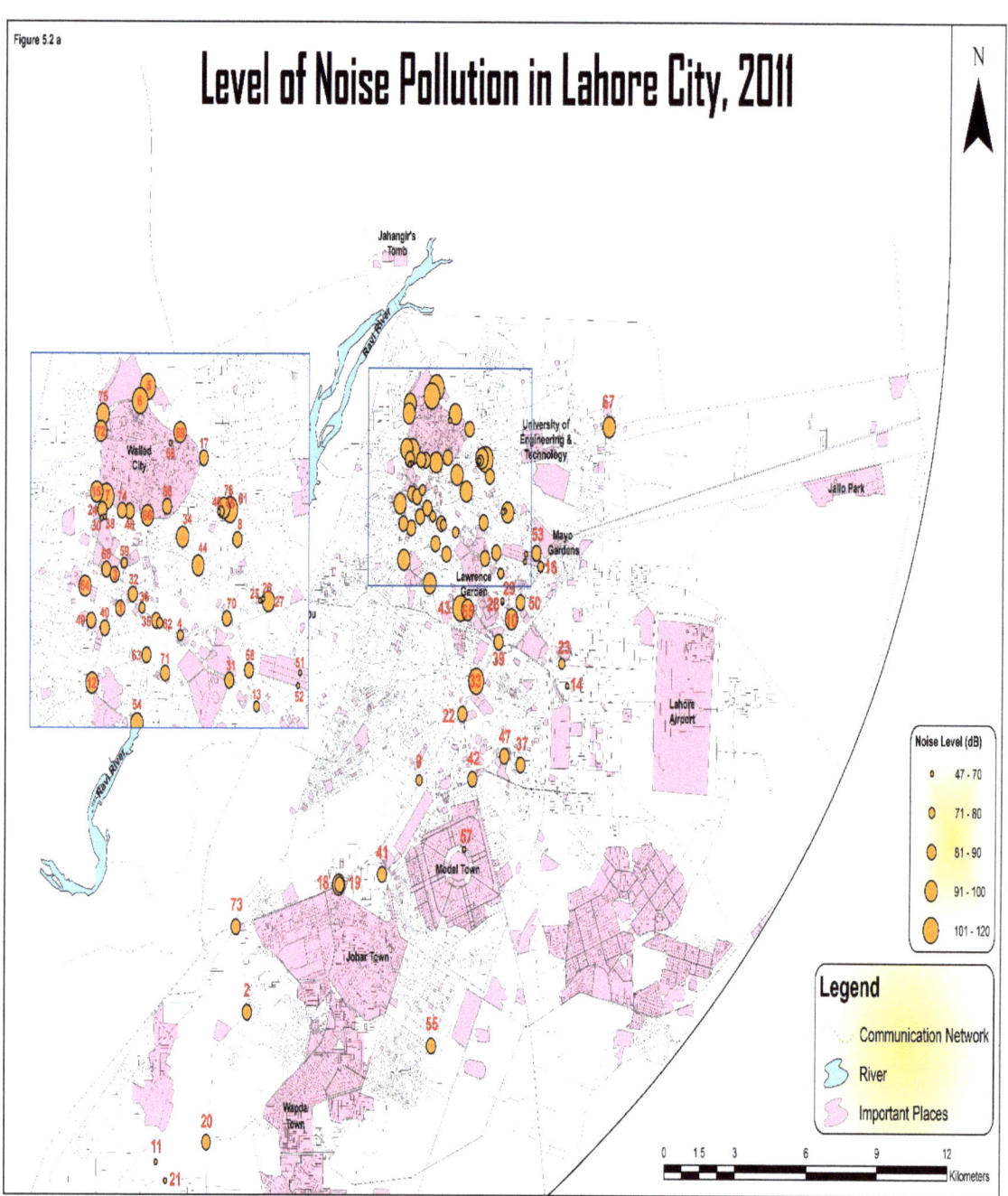

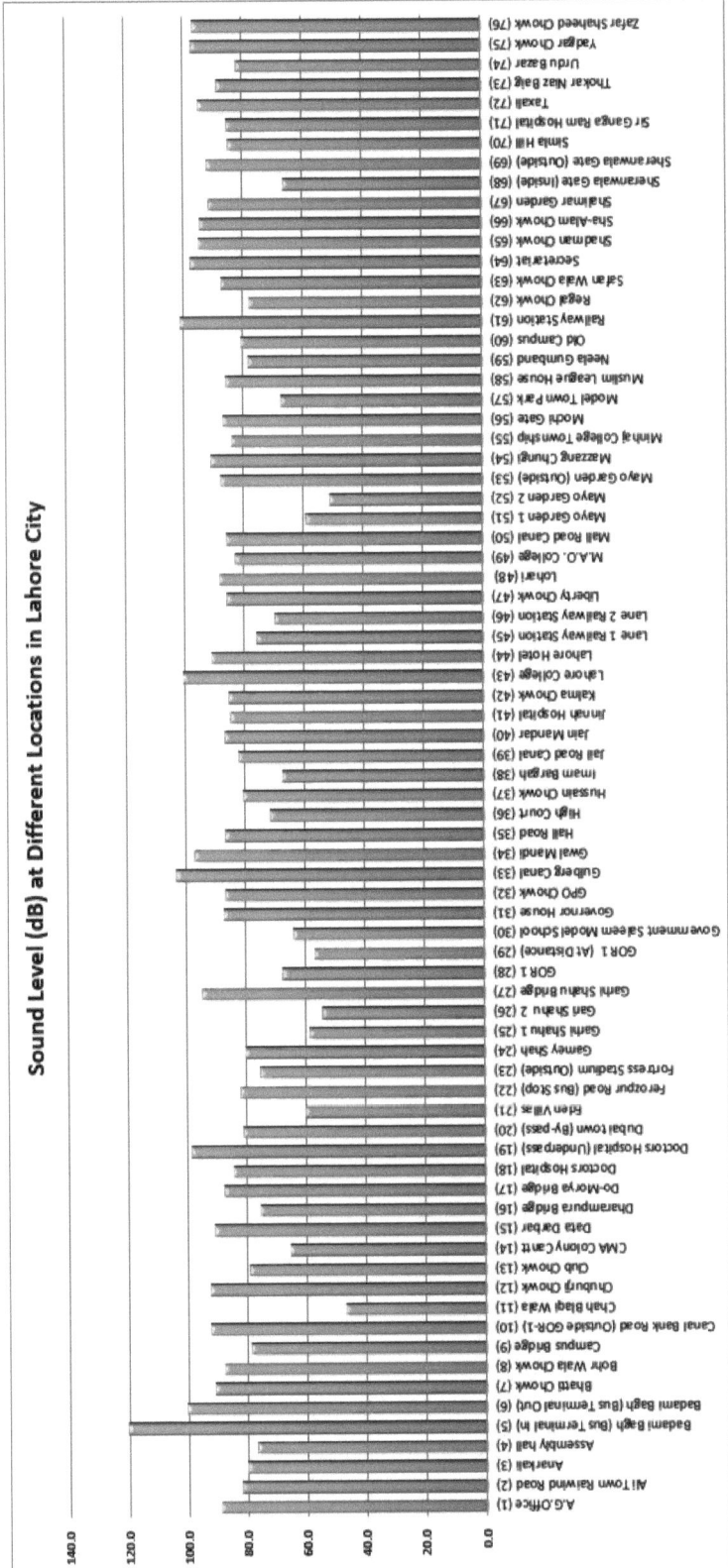

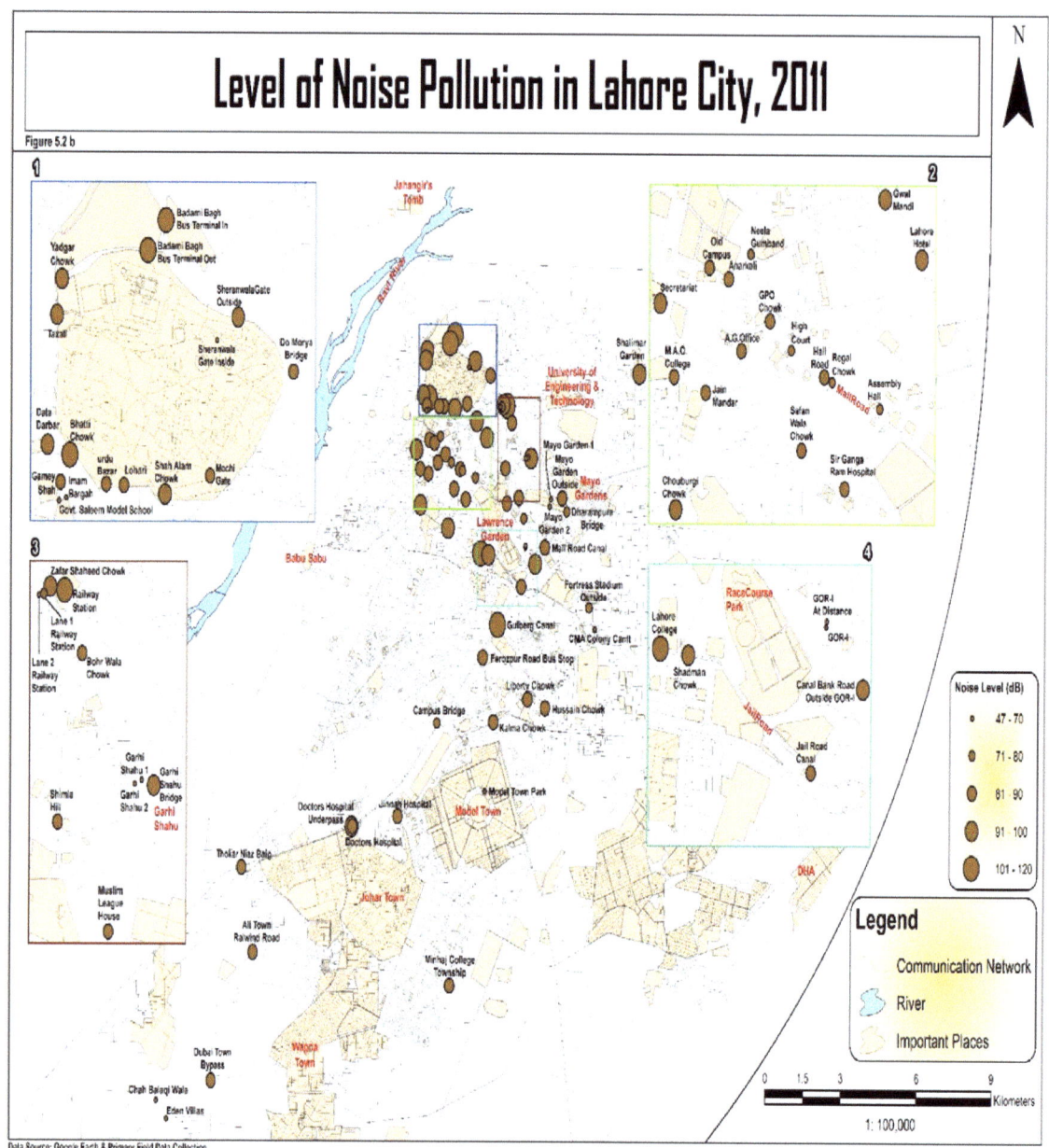

Figure 5.2 b

5.2 Level of Noise Pollution around Walled City Lahore, 2011

Figure 5.3 explains the level of noise pollution in and around Walled city. Badami Bagh was found to be the noisiest place and the cause of noise was unplanned vehicular traffic. It is a bus terminal for Lahore city but has narrow roads, which lead to heavy traffic flow even at any time of the day. The level of noise inside the bus terminal as well as outside the bus terminal was very high and it was exceeding the permissible limits. Bhatti Chowk was another noisy place where noise level was 91 dB (A). The duration of exposure to noise is more than 12 hours at these places especially at Badami Bagh. Noise pollution at Yadgar Chowk, Taxali, outside Sheranwala Gate and Shah Alam Chowk falls between 91-100dB (A) and it exceeds the permissible limits. Yadgar Chowk, Ali Hajvery Darbar road and Taxali have heavy traffic flow but have wide roads due to which the load of traffic becomes less. At Sheranwala gate the volume of traffic is high and the road is narrow. Another important reason of noise in these areas is that the traffic entering the city passes through these points and then gets diverted to different roads. Shah Alam Chowk is another noisy place. It leads to Shah Alam Market, which is the Central Business District of the city. The major causes of noise in this area are the business activity, narrow roads and load of light traffic that get entangled at many points. Ali Hajvery Darbar and Taxali chowk are also busy points where roads are although wide but load of traffic is very high. Mochi gate, Lohari, Urdu Bazar, Gamey Shah and Do Morya bridge are the points where noise level ranges from 81-90 dB (A). The road network of these points is narrow and traffic load is very high. Motor cycle rickshaw is considered to be the means of movement for local population. The major source of noise out of all the motor vehicles was the motor cycle rickshaw. Roads of Urdu Bazar are narrow. There are two more locations that of Govt. Saleem Model School and Imam Bargah where noise levels are 64dB (A) and 67.6 dB (A) successively. The major reason of low noise levels is that these locations were taken at a distance of 100 meter away from the road.

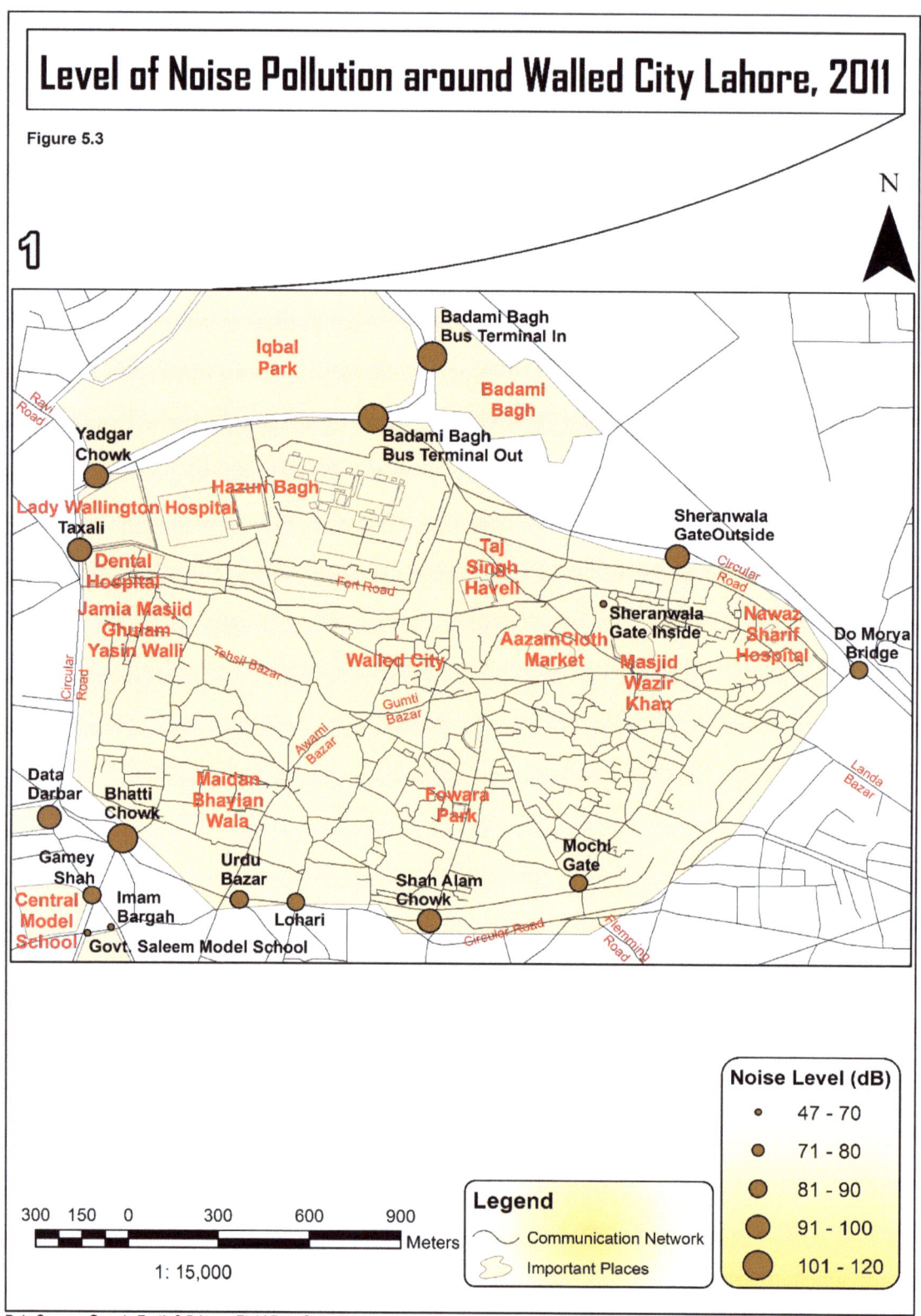

Table 5.1: Level of Noise Pollution around Walled City Lahore, 2011

Id	Location	Sound Level dB (A) Average
1	Badami Bagh (Bus Terminal In	120.4
2	Badami Bagh (Bus Terminal Out	100.7
3	Bhatti Chowk	91.1
4	Data Darbar	90.9
5	Do-Morya Bridge	87.6
6	Gamey Shah	80.3
7	Government Saleem Model School	64.0
8	Gwal Mandi	97.1
9	Imam Bargah	67.6
10	Lohari	88.4
11	Sha-Alam Chowk	94.5
12	Sheranwala Gate (Inside)	66.4
13	Sheranwala Gate (Outside)	92.1
14	Taxali	94.9
15	Yadgar Chowk	97.1

Source: Primary data collected through SLM.

5.3 Level of Noise Pollution around Mall Road Lahore, 2011

Figure 5.4 shows level of noise pollution along Mall road Lahore. The noisiest places in this map are the Secretariat Lahore and Chouburgi Chowk, Gwal Mandi and Lahore Hotel where noise levels were 97.6 dB (A) and 92.7 dB (A), 97.1 dB (A) and 91.3 dB (A) successively.

The main cause of high level of noise pollution in these areas is that the roads along these points fetch maximum number of vehicles, which are then diverted to different directions. Sir Ganga Ram Hospital, Hall Road, GPO Chowk, Anarkali, Old Campus, A.G. Office, Jan Mandar and Safan Wala Chowk are the points where noise levels ranges from 81-90 dB (A). Sir Ganga Ram is a hospital where the noise level must not exceed 50 dB (A) to 55 dB (A).

The noise level exceeds the permissible limits. Another sensitive point is Old Campus of the Punjab University, Lahore. At this location the noise level should follow the noise level of silence zone but actual values exceeds the permissible limits.

GPO Chowk, Anarkali, Jan Mandar and Safan Wala Chowk are the busiest places where volume of traffic is high and roads are narrow.

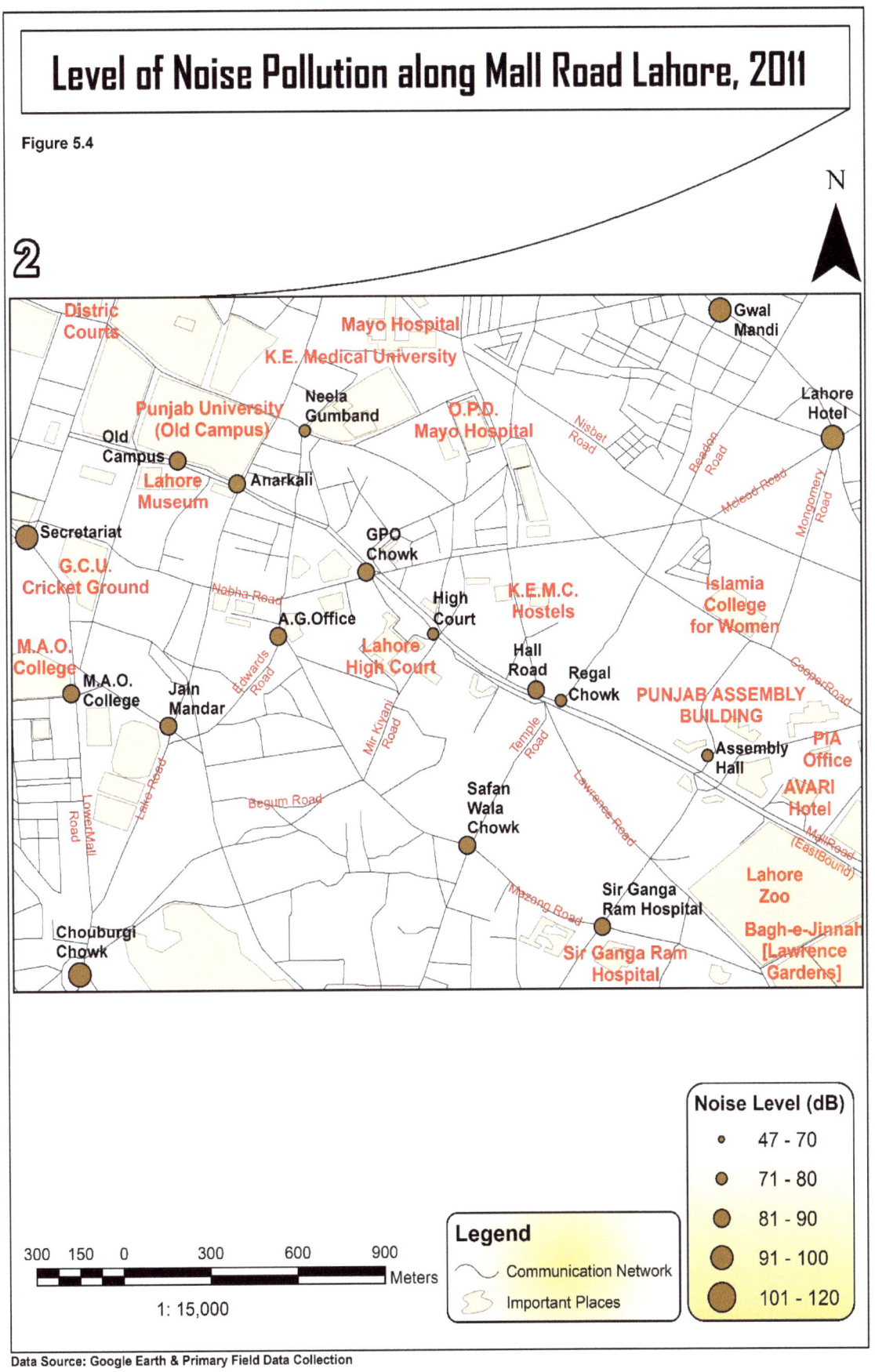
Figure 5.4: Level of Noise Pollution along Mall Road Lahore, 2011

Table 5.2: Level of Noise Pollution around Mall Road Lahore, 2011

Id	Location	Sound Level dB (A) Average
1	A.G.Office	89.4
2	Anarkali	80.2
3	Assembly hall	76.7
4	Chuburji Chowk	92.7
5	GPO Chowk	86.9
6	Gwal Mandi	97.1
7	Hall Road	86.7
8	High Court	71.7
9	Lahore Hotel	91.3
10	Mall Road Canal	85.9
11	Neela Gumband	78.3
12	Old Campus	80.8
13	Regal Chowk	78.0
14	Safan Wala Chowk	87.4
15	Secretariat	97.6
16	Sir Ganga Ram Hospital	85.4

Source: Primary data collected through SLM.

5.4 Level of Noise Pollution around Jail Road Lahore, 2011

Figure 5.5 shows many important locations of the City along Jail road Lahore. It is one of the busiest roads. Jail road connects Ferozpur road with Canal road. This road touches many important hospitals and colleges. The noise levels along Jail road have increased the maximum permissible limits set by Environment Protection Department. The noise level must not exceeds 55 dB (A) during the day time and 45 dB (A) during the night time. The maximum noise value was recorded in front of Lahore College, which is 1000.9 dB (A).

The noise level was also high at Shadman Chowk and at canal bank outside GOR-1. At Shadman chowk it was 95.1 dB (A) and 92.6 dB (A) successively. The major cause of high noise level at above discussed locations is high volume of traffic. These roads have the importance of jugular vein for the City. As far as Jail road canal is concerned the main cause of noise is the narrow road and high volume of traffic. The important thing is the difference in noise values inside GOR-I and outside along the Primary road. Inside GOR-I the noise level is 57.1 dB (A), which is very low as compared to the noise level outside. There are two reasons of reduction in noise pollution at these locations. First of all it is not a thoroughfare. Secondly the green belt inside GOR-I also absorb noise coming from outside.

Table 5.3: Level of Noise Pollution around Jail Road Lahore, 2011

Id	Location	Sound Level dB (A) Average
1	Canal Bank Road (Outside GOR-1)	92.6
2	GOR 1 (At Distance)	57.1
3	GOR 1	67.6
4	Jail Road Canal	82.1
5	Lahore College	100.9
6	Shadman Chowk	95.1

Source: Primary data collected through SLM.

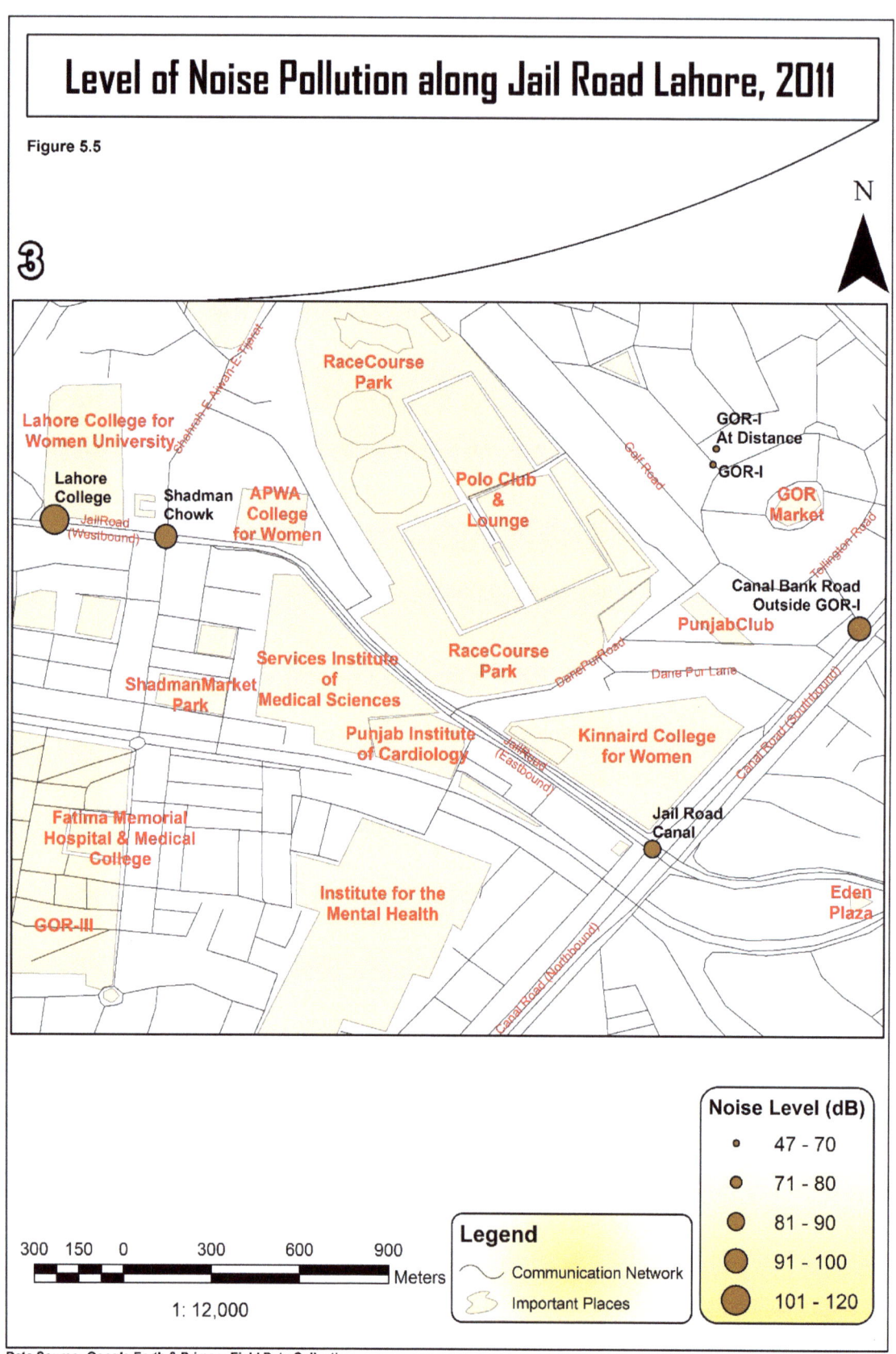
Figure 5.5 Level of Noise Pollution along Jail Road Lahore, 2011

5.5 Level of Noise Pollution around Garhi Shahu Lahore, 2011

Figure 5.6 shows that Railway Station is the most polluted sample site where noise level is 101.2 dB (A). The main cause of noise at Lahore Railway station is not only the railway but it is also a junction of intercity transport. The roads are narrow but the traffic volume is high. Motor cycle rickshaw is considered to be the major cause of noise. Traffic blockade is a routine matter at this point because traffic from different directions of the city comes finally at this point.

There are two more sample sites other than railway station where the noise level ranges between the railway en 91-100 dB (A). These points are Zafar Shaheed Chowk and Garhi Shahu Bridge. Zafar Shaheed Chowk. All these points are nearest to the railway station. There location gives logic to the high noise level.

Garhi Shahu is a midpoint between railway station and Canal Road. It is a busy chowk with many business activities and high traffic volume. Traffic passes through this point towards different directions. The road is also narrow, which add fuel to fire. Traffic jams lead to the use of horns.

All the factors collectively make this point difficult to cross. As we move away fro the main road toward Mayo Garden one can see a marked decrease in noise levels, which is mainly due to green belts. There are three locations where noise level ranges between 81-90 dB (A). These points are Muslim League House, Simla Hill and Bohr Wala Chowk. At all these points the roads are narrow.

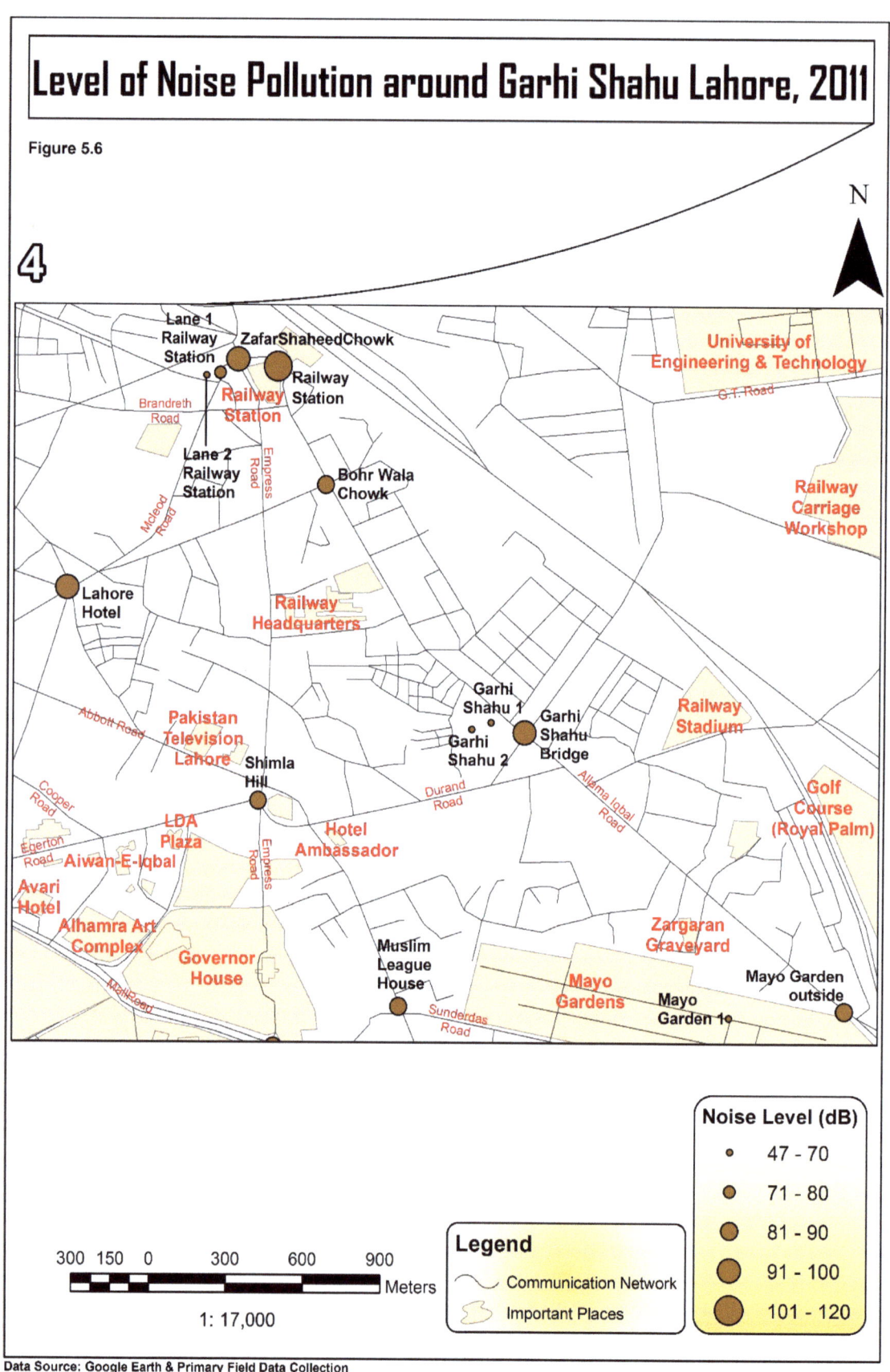

Figure 5.6

Table 5.4: Level of Noise Pollution around Garhi Shahu Lahore, 2011

Id	Location	Sound Level dB (A) Average
1	Bohr Wala Chowk	87.9
2	Garhi Shahu 1	58.7
3	Garhi Shahu Bridge	94.8
4	Gari Shahu 2	54.3
5	Lane 1 Railway Station	75.7
6	Lane 2 Railway Station	69.7
7	Mayo Garden (Outside)	87.8
8	Mayo Garden 1	59.1
9	Mayo Garden 2	51.1
10	Muslim League House	86.0
11	Railway Station	101.2
12	Shimla Hill	84.8
13	Zafar Shaheed Chowk	96.6

Source: Primary data collected through SLM.

5.6 Area wise Coverage of Noise Pollution

Figure 5.7 shows the extent of noise around the source area. Source is the location where a motor vehicle generates some noise. The GIS technique used to show the extent of noise is the creation of buffer zones, which give proximity analysis. Buffers are created according to the intensity of sound produced by a motor vehicle. The high intensity buffer areas have been shown in pink shade showing radius of 200 meters and the area surrounded by each buffer shows that the people living within this zone directly suffers from the noise related problems.

Then comes medium intensity buffers in yellow color are showing 150 meter radius. These buffers show noise level range from 71-90 dB (A). The cyan blue color of buffers shows the low intensities of noise. These buffers show noise level range with a radius of 50 meters. Figure 5.7 highlight the extent of areas covered by buffer where noises coming from different directions disturb the normal activities of man.

In and around walled city the noise level is high as well as roads are narrow as compared to the need of traffic volume. Absence of green belt throughout the area makes it noisier. People living in walled city are often annoyed by noise coming from outside source. External noise from emergency vehicles, traffic, construction work and other city noises create problem for the dwellers. Adjacent vertical buildings in narrow streets along the major roads make their life horrible. Exposure to excessively loud sounds over long periods has badly affected the hearing of people in these areas. In these areas 56% residents are suffering from Temporary hearing loss. Noise causes anxiety and tension in the dwellers. The sample points along Ferozpur road also show buffers of 200 meter radius, which means that the annoyance rate is high among people at these locations. Along Mall road the buffers with 150 meter radius show medium range of noise. The reason is that the road is wide as compared to roads of walled city and the traffic flow is also less and properly planned. A ban of motorcycle rickshaw on the Mall road also helps reducing the noise levels. The buffer zones in Gulberg. Thokar Niaz Beig and Township show medium intensity of noise. The areas of Chah Blaqi Wala and Eden Villas show low intensity of noise. All the areas showing buffer zone with a radius of 200 meter are extremely high risk zones. The areas with 150 meter radius buffer are high risk zones and areas with 50 meter buffer radius are medium risk zones. People in high risk zone areas are more prone to noise induced hearing loss.

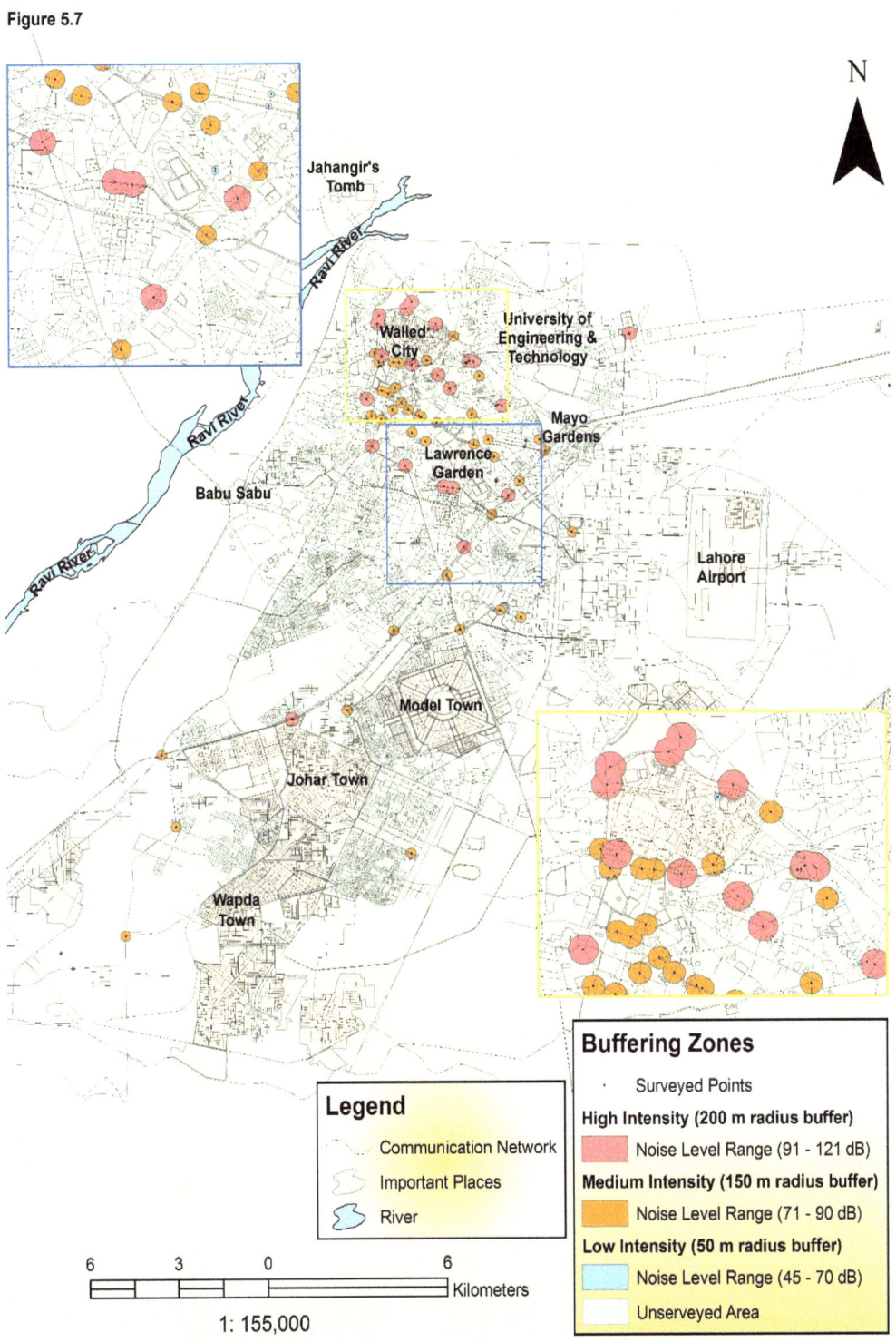

Figure 5.7

5.7 Comparative analysis of morning and evening levels

The comparative analysis of noise pollution during morning and evening gives a clear picture of daily changes in noise levels. During the morning time the noise levels were high at Shalamar Chowk, Badami Bagh inside and outside, Yadgar Chowk, Bhatti gate and Railway Station around walled city. It is also very high at, Shadman Chowk and Along Jail road canal bank as shown in the figure 5.8.a. The high values of noise have been shown in map with red color, which means that noise level at all these places ranges from 91-121 dB (A). The reasons of high noise levels are the high volume of traffic, use of motorcycle rickshaw, undisciplined traffic and narrow roads etc. The people living in these areas are suffering from severe problems related to hearing. The cyan color highlights the areas of medium noise intensity with a range of 71-90 dB (A). Doctors hospital, Jinnah hospital, Assembly Hall, Bahtti chowk, Bohr wala chowk, Ali Hajvery Darbar, Ferozpur road bus stop, Dubai town (By-pass), Gamey Shah, GPO, Hall road, Hussain Chowk, ain Mandar, Kalma Chowk, Lohari, Liberty Chowk, Mayo Garden, Mozang Chungi, Safan Wala Chowk, Sheranwala gate, Dharampura Bridge, Sir Ganga Ram hospital and Urdu Bazar are the places where noise intensity ranges from 71-90 dB (A). The high level of noise in these areas is mainly due to heavy volume of traffic. The third category consists of those places where noise intensity ranges between 45-70 dB (A). Such places are Regal Chowk, High Court, Imam Bargah, Railway station lane 1, Mayo Garden 1 and 2, Chah Blaqi Wala, CMA Colony Cantt, Fortress stadium, Garhi Shahu 1 and 2 and Govt. Saleem Model School. These are the places, which are either away from the major roads or have green belt around them. People living in these areas are not suffering from hearing problems as frequent as the people living in high range noise level areas. Figure 5.8.b. shows that in the evening noise pollution decreases at some places such as Badami Bagh outside, Dharampura Bridge and Sheranwala gate etc. This decrease in noise level is due to decrease in the traffic volume. Even then the values of noise pollution cross the maximum permissible limits. Figure 5.8.c shows a comparative analysis of both that timings collectively.

Table 5.5: Morning and evening noise levels

Serial No.	Location	Morning	Evening
1	A.G.Office	85	88
2	Ali Town Raiwind Road	82	80
3	Anarkali	79.5	78
4	Assembly hall	73	79
5	Badami Bagh (Bus Terminal In)	115	121.3
6	Badami Bagh (Bus Terminal Out)	103	89
7	Bhatti Chowk	87	88
8	Bohr Wala Chowk	87	83
9	Campus Bridge	80	76
10	Canal Bank Road (Outside GOR-1)	88.4	98
11	Chah Blaqi Wala	48	44
12	Chuburji Chowk	90	92.2
13	Club Chowk	76	80
14	CMA Colony Cantt	60	66
15	Data Darbar	88	90
16	Dharampura Bridge	70.6	68.1
17	Do-Morya Bridge	87	84
18	Doctors Hospital	86	83
19	Doctors Hospital (Underpass)	95	99
20	Dubai town (By-pass)	87	75
21	Eden Villas	60	59
22	Ferozpur Road (Bus Stop)	85	79.7
23	Fortress Stadium (Outside)	70	81
24	Gamey Shah	75	80
25	Garhi Shahu 1	50	60
26	Gari Shahu 2	55	51
27	Garhi Shahu Bridge	99	89
28	GOR 1	65	67
29	GOR 1 (At Distance)	57	52
30	Government Saleem Model School	56	70
31	Governor House	88.5	82.6
32	GPO Chowk	88	83.5
33	Gulberg Canal	100	98
34	Gwal Mandi	100	89
35	Hall Road	86	81.7

36	High Court	67	71
37	Hussain Chowk	77	80.3
38	Imam Bargah	65	67
39	Jail Road Canal	82	77
40	Jain Mandar	80	90
41	Jinnah Hospital	85	85
42	Kalma Chowk	80	89
43	Lahore College	95	99.2
44	Lahore Hotel	90	93
45	Lane 1 Railway Station	70	78
46	Lane 2 Railway Station	67	70
47	Liberty Chowk	83	89
48	Lohari	89	84
49	M.A.O. College	80	81
50	Mall Road Canal	85	85
51	Mayo Garden 1	59	58
52	Mayo Garden 2	53	45
53	Mayo Garden (Outside)	85	88
54	Mazzang Chungi	88	90
55	Minhaj College Township	90	74.4
56	Mochi Gate	90	77
57	Model Town Park	65	67
58	Muslim League House	90	78
59	Neela Gumband	77	76
60	Old Campus	80.5	77
61	Railway Station	98	88.1
62	Regal Chowk	70	80
63	Safan Wala Chowk	80	90
64	Secretariat	95	97
65	Shadman Chowk	99	90
66	Sha-Alam Chowk	85	100
67	Shalimar Garden	97.5	87
68	Sheranwala Gate (Inside)	66	65
69	Sheranwala Gate (Outside)	85	98
70	Simla Hill	80	87
71	Sir Ganga Ram Hospital	83	88.3
72	Taxali	99	87
73	Thokar Niaz Baig	80	99.4
74	Urdu Bazar	80	72
75	Yadgar Chowk	102	88
76	Zafar Shaheed Chowk	99	88

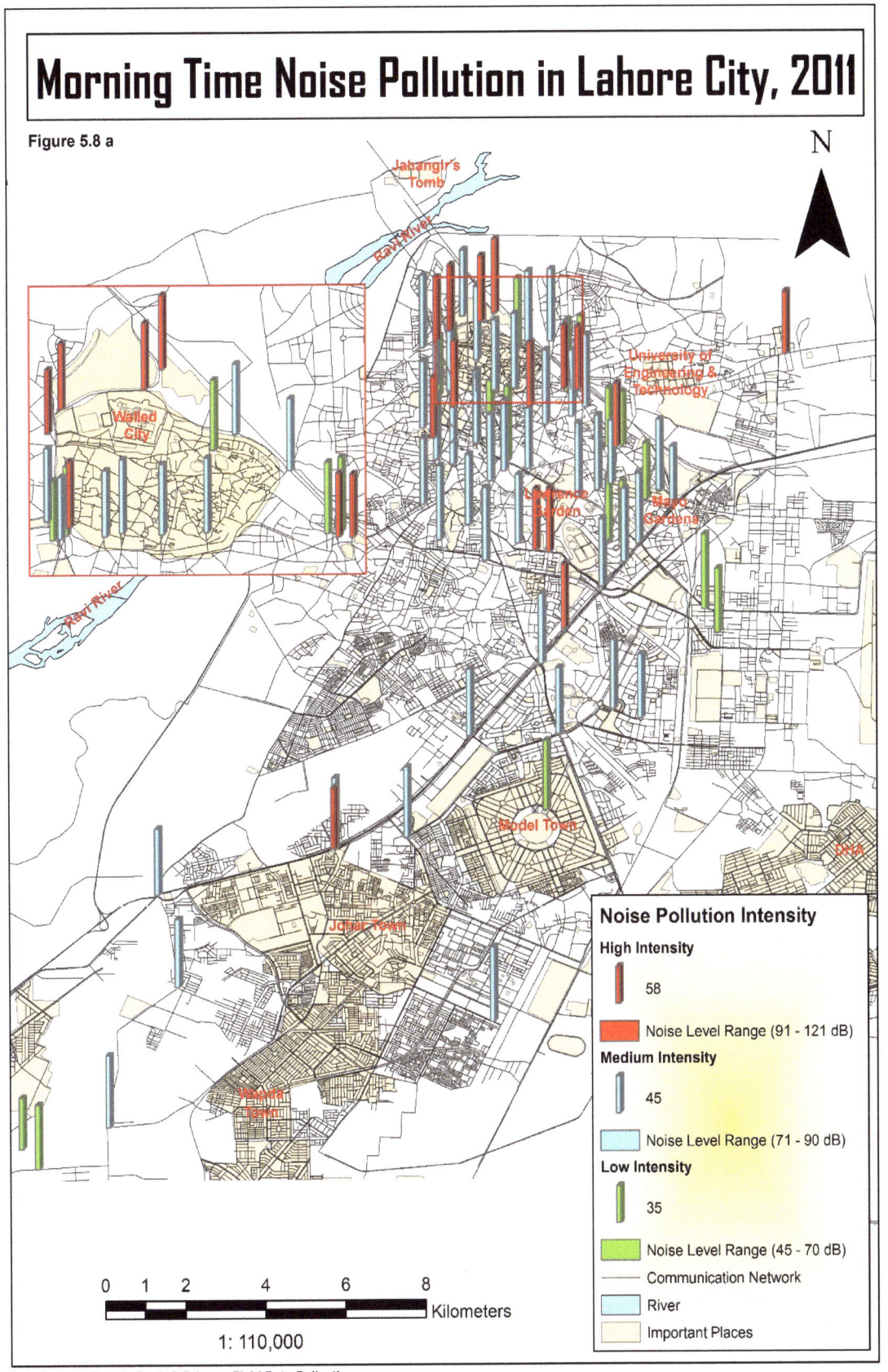

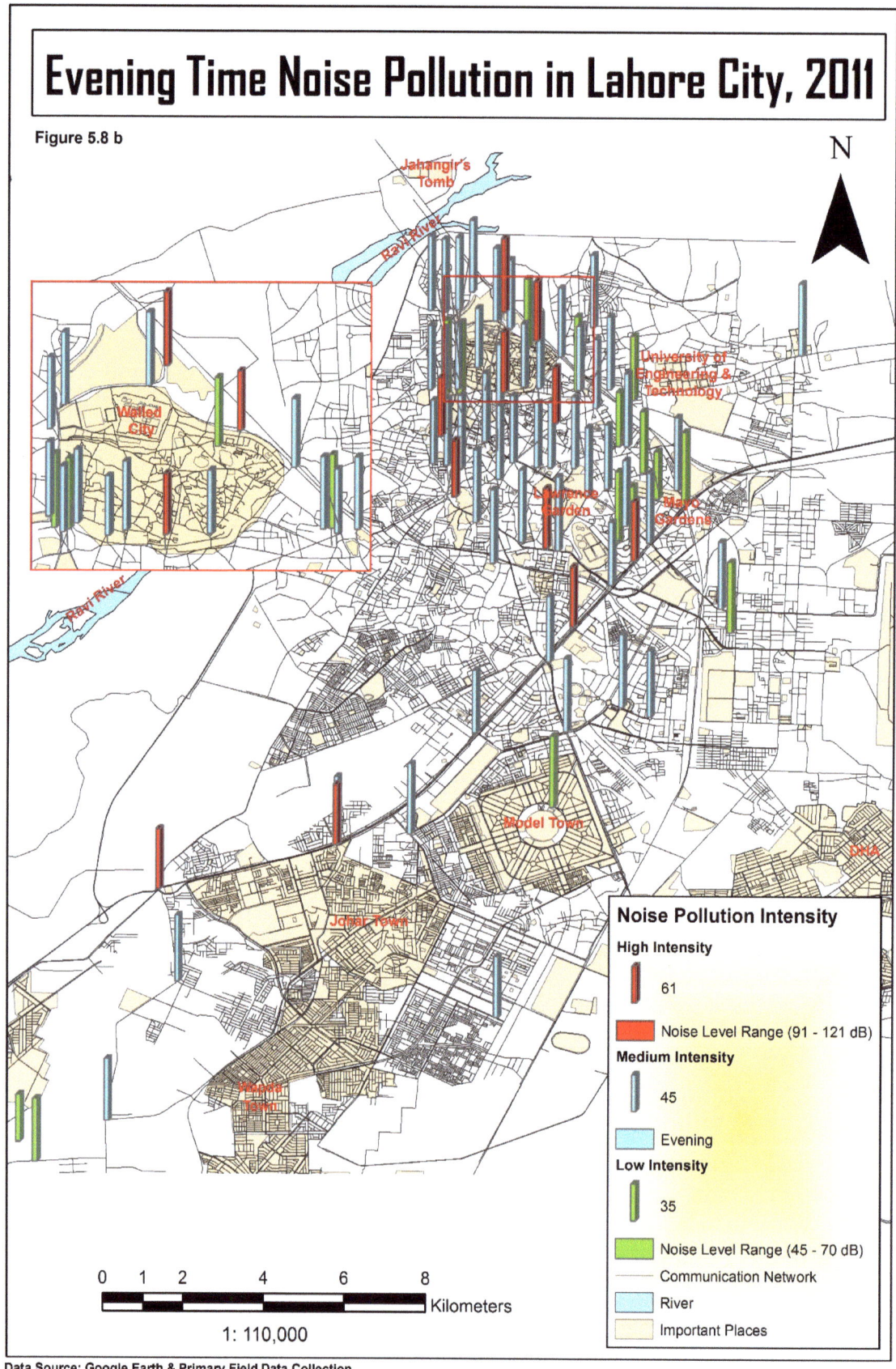

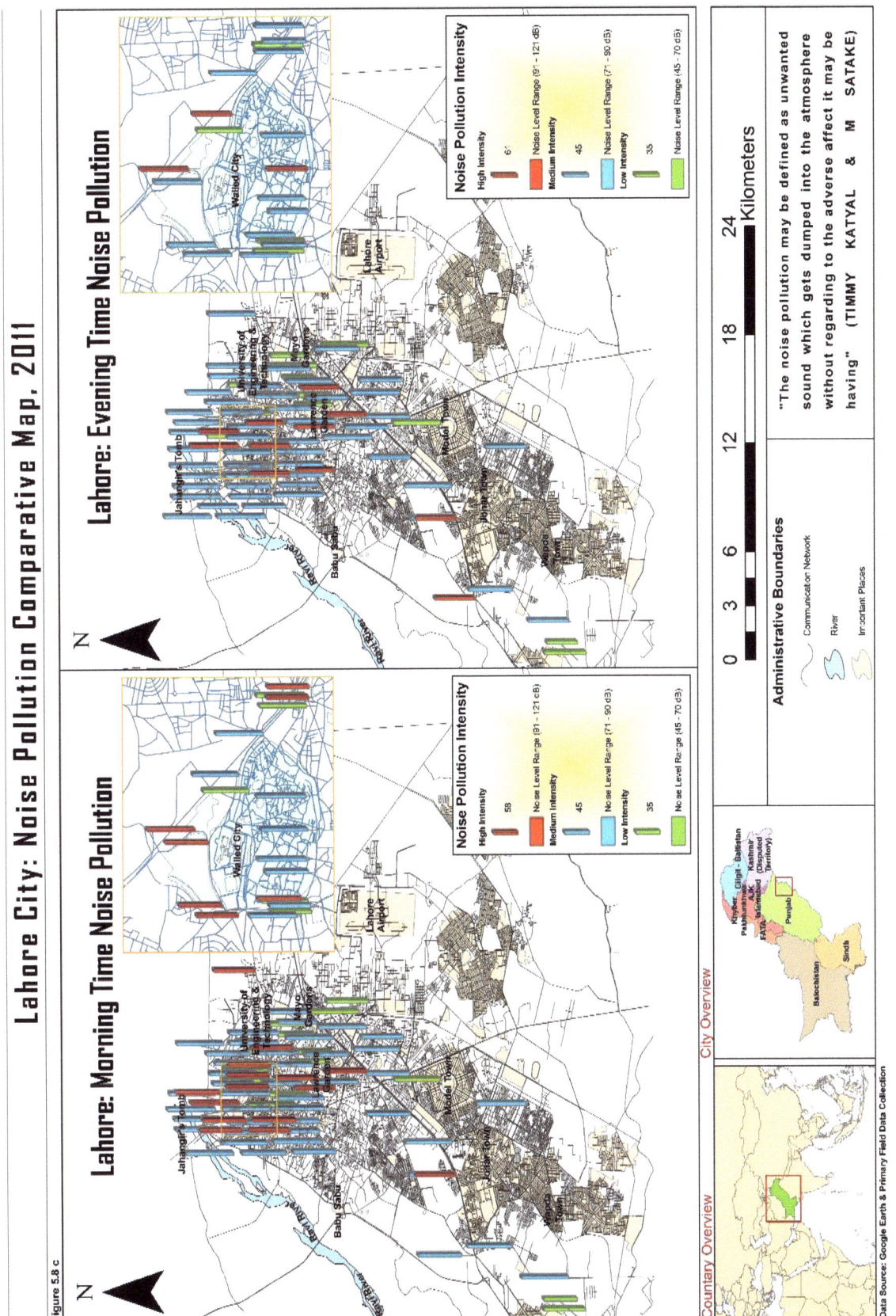

5.8 Statistical Analysis

A preliminary survey was carried out among 125 inhabitants/workers by questionnaire method to gather information about the suffering of noise related health problems. The survey was conducted at those locations where the level of noise was high. These were eight sample points such as Badami Bagh inside, Taxali, Data Darbar, Sheranwala Gate, Infront of Shalimar Gardena, Railway Station, Garhi Shahu and Thokar Niaz Baig as shown in the figure 5.9.

The questionnaire survey was also conducted at two such places where the level of noise was within the permissible limits. These locations were Campus Bridge and Chah Blaqi Wala as shown in figure 5.10.

The Questionnaire was consisting of twenty three questions. All these questions were regarding noise pollution. There were three kinds of questions. A few questions were about causes of noise pollution. A few of them were about effects of noise pollution and rests of them were about suggestions for improvement.

In the first question it was asked that do you think that traffic density is the root-cause of increasing noise pollution. 100% respondents agreed with the cause. The second question was that is population one of the major causes of increase in number of vehicles and noise pollution. Most of the people disagree with this question because they were of the view that if the law enforcing authorities improved the infrastructure and maintain traffic discipline the increase in population alone is not a major cause. The suggestions were given that the use of public buses for massive transport instead of personal cars can be encouraged, which will help bringing the noise levels down.

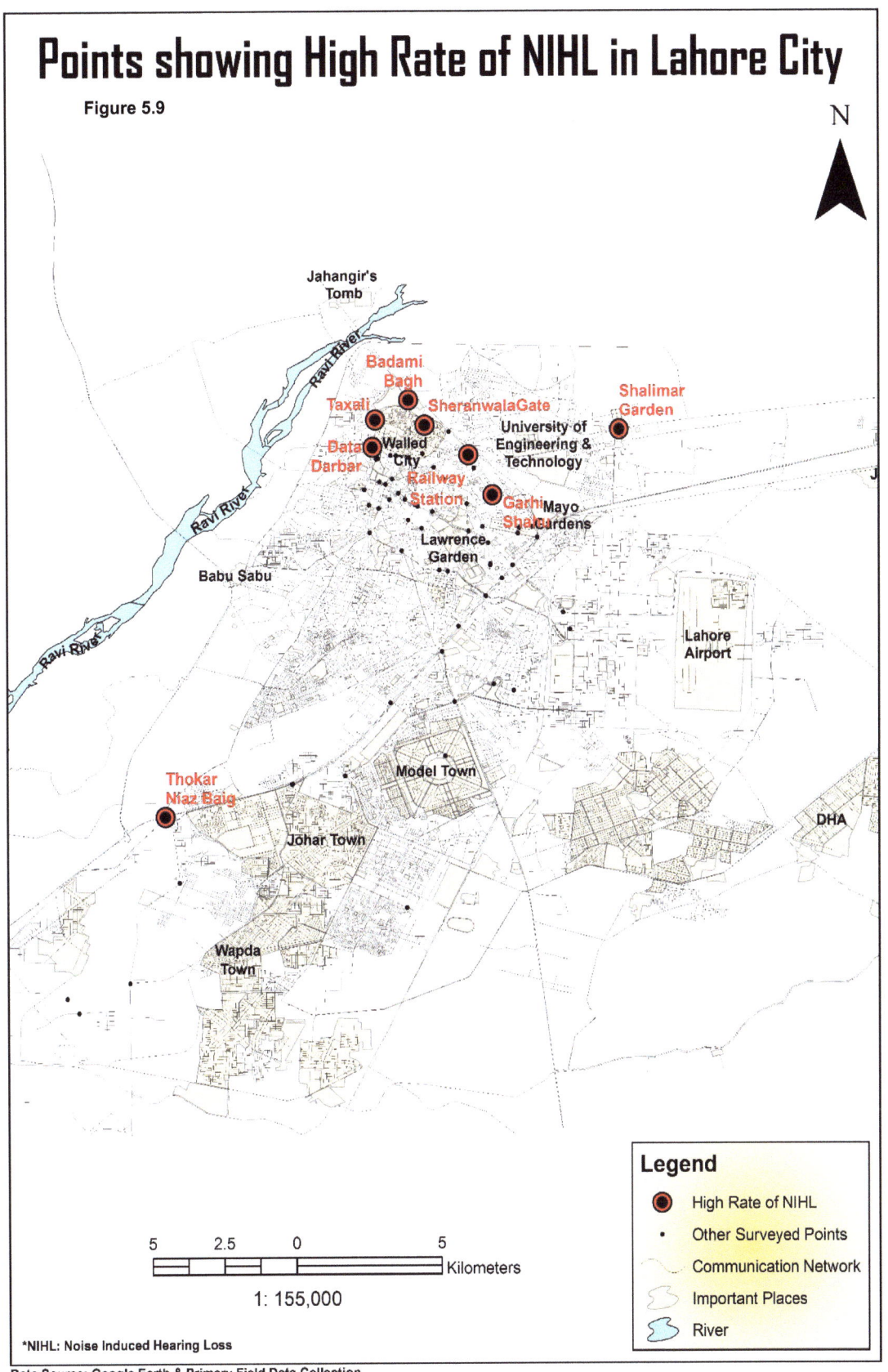

Figure 5.9 Points showing High Rate of NIHL in Lahore City

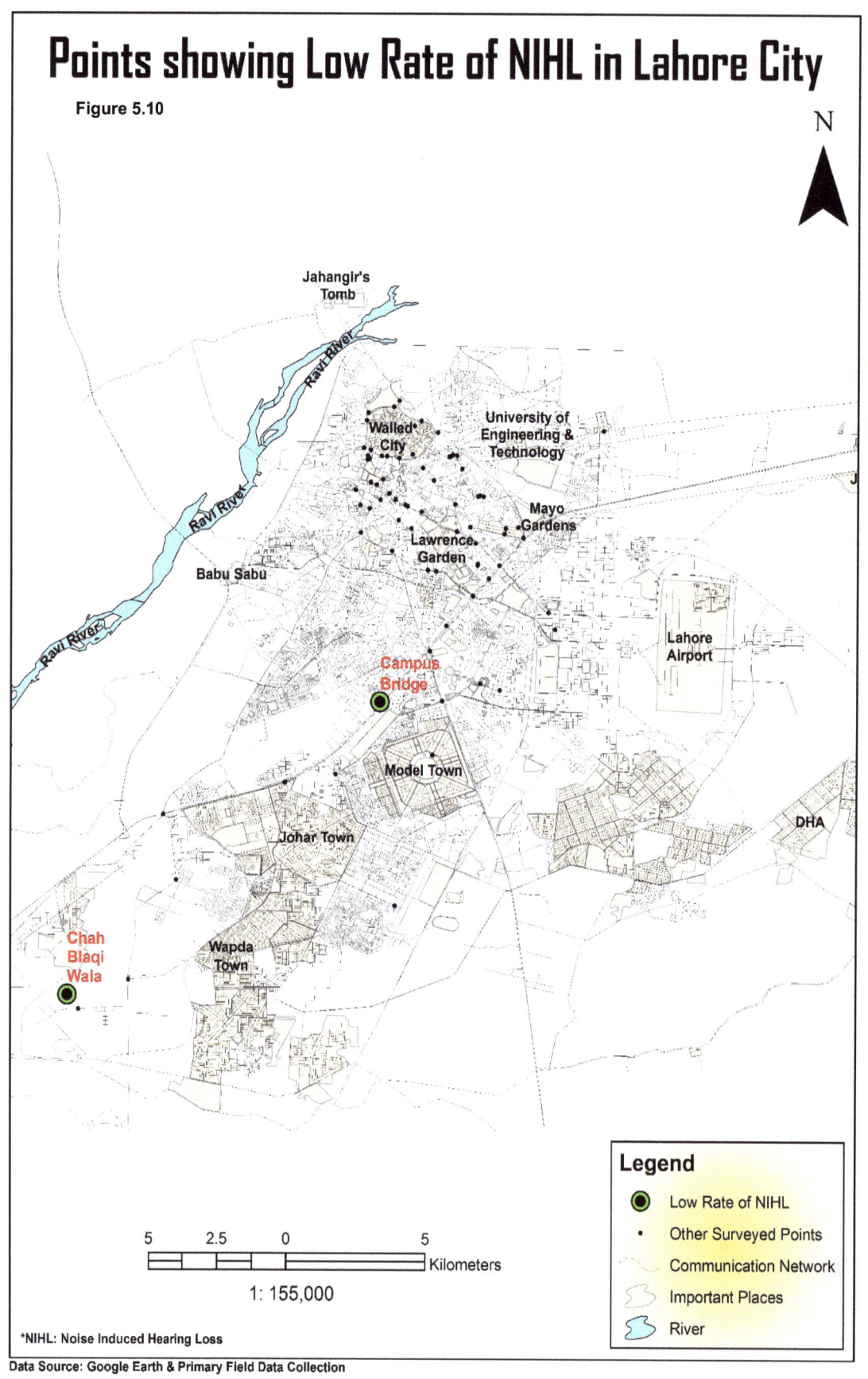

Figure 5.10: Points showing Low Rate of NIHL in Lahore City

The question number three was regarding the noise pollution due to penetration of industries/ factories in the non-commercial areas. In reply to this question most of the respondents disagree. According to them the problem of noise pollution regarding set up of industries is prominent only at a few locations. The major cause of noise pollution in case of the City is not industry but it is vehicular traffic.

The question number four was about business activities and noise pollution. The respondents at most of the places replied positively especially in walled city, Garhi Shahu, Shalamar Chowk and Thokar Niaz Baig where business activities attract people as well as more traffic.

Next question was about the awareness about traffic laws. Approximately 50% respondents were aware about these laws but rests of the respondents were not aware of traffic laws. This point of questionnaire highlights the need of awareness campaigns among masses by government agencies.

Most of the roads of the city are in bad condition that is why it was also made part of questionnaire. It was asked that whether you think that complex and narrow road networks are the reasons of increasing traffic density. The answers were yes in most of the areas of Walled city, Garhi Shahu and Railway Station. Most of the respondents suggested that wide network of roads can make traffic flow easy, which will in turn lower the use of horns. Another question asked was do you think that there is lack of traffic discipline, which causes more noise pollution. Most of the respondents replied positive and complaint against the traffic police, which prefer bribe than discipline.

The next portion of questionnaire was about the effects of noise pollution. In the first question of this potion it was asked that do you think that the diseases like frustration and stress are caused by noise pollution. All the respondents strongly agreed with it. It was asked in the next question that are you suffering from hearing loss? 90% of respondents replied that they are suffering from temporary hearing loss, which shows that hearing loss is emerging as a major threat in most of the areas of the City.

It was also asked that is it cause of depression, tension, hearing loss, high blood pressure, insomnia and lack of concentration on the work for you. The answers were yes, which reflect that people are suffering from many diseases due to noisy culprit.

The third part of the questionnaire was about the suggestions. In this portion many questions were asked about the need of mature road network, role of EPA and TEPA, Self-learning and awareness campaigns by government, Making hospitals and schools noise free zones and strict action against law breakers. 100% respondents strongly agreed with these points.

Respondents gave many suggestions especially regarding widening of roads, discipline of the traffic and improvement in technology used in the silencer of motor cycle rickshaws.

After getting the results of the survey the data were arranged in tabular form and a statistical analysis was done to see the relationship among variables. For this purpose software SPSS was used. The analysis chosen was the hierarchical cluster analysis, which presented the data in the form of dendogram.

5.8.1 Cluster analysis

It classifies a set of observations into two or more mutually exclusive unknown groups based on combinations of interval variables. The purpose of cluster analysis is to discover a system of organizing observations, usually people, into groups where members of the groups share properties in common. It is cognitively easier for people to predict behavior or properties of people or objects based on group membership, all of whom share similar properties (http://en.wikipedia.org/wiki/Cluster_analysis).

The trick in cluster analysis is to collect information and combine it in ways that allow classification into useful groups. The hierarchical cluster analysis was applied on noise pollution data of city of Wenzhou and was concluded that the main noise sources of the city were social life noise and industrial production noise (QiWang, 2008).

Cluster analysis starts with a data matrix, where objects (usually people in the social sciences) are rows and observations are columns.

The difference between a proximities matrix in cluster analysis and a correlation matrix is that a correlation matrix contains similarities between variables while the proximities matrix contains similarities between observations.

A dendogram that clearly differentiates groups of objects will have small distances in the far branches of the tree and large differences in the near branches. When the distances on the far branches are large relative to the near branches, then the grouping is not very effective.

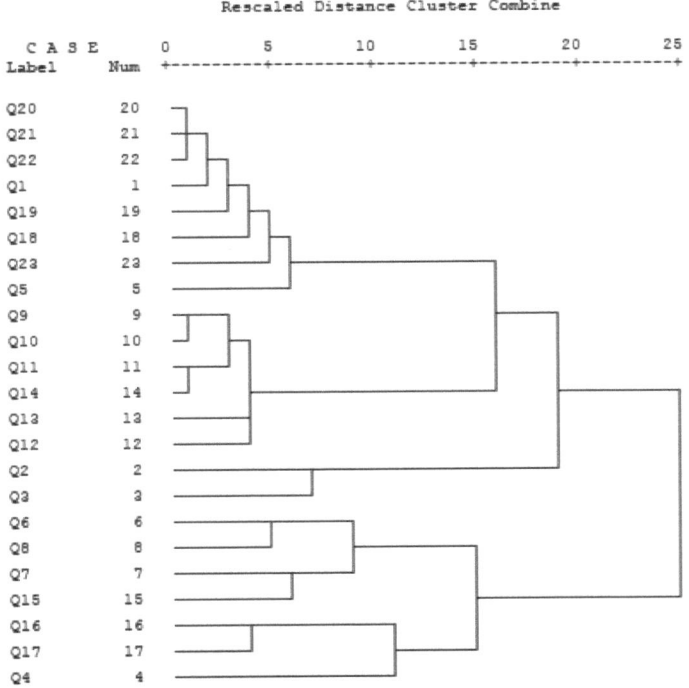

Figure 5.11:Dendogram

Table 5.6: Case Processing Summary (Squared Euclidean Distance used)

Cases					
Valid		Missing		Total	
N	Percent	N	Percent	N	Percent
125	100.0%	0	.0%	125	100.0%

This dendogram shows that respondents strongly agreed that government needs to take strict action against law breakers to implement traffic rules as well as environment protection laws. They also agreed strongly that a threshold limit of noise levels should also be fixed around these hospitals and educational institutions.

Table 5.7: Agglomeration schedule

Stage	Cluster Combined		Coefficients	Stage Cluster First Appears		Next Stage
	Cluster 1	Cluster 2	Cluster 1	Cluster 2	Cluster 1	Cluster 2
1	20	21	39.000	0	0	4
2	9	10	49.000	0	0	7
3	11	14	56.000	0	0	7
4	20	22	57.500	1	0	5
5	1	20	80.333	0	4	6
6	1	19	85.250	5	0	8
7	9	11	94.000	2	3	9
8	1	18	101.800	6	0	13
9	9	13	103.250	7	0	11
10	16	17	108.000	0	0	18
11	9	12	118.800	9	0	20
12	6	8	142.000	0	0	17
13	1	23	142.500	8	0	15
14	7	15	147.000	0	0	17
15	1	5	156.857	13	0	20
16	2	3	171.000	0	0	21
17	6	7	219.500	12	14	19
18	4	16	257.000	0	10	19
19	4	6	332.917	18	17	22
20	1	9	360.208	15	11	21
21	1	2	419.643	20	16	22
22	1	4	558.884	21	19	0

There is a strong relationship between the traffic density and high noise levels. The figure 5.11 explains that the density of traffic needs to be reduced in areas of hospitals and educational institutions and they should be declared as noise free zones. Figure also explains that a campaign can be introduced by the government to reduce noise pollution.

The figure also explains that self- learning and awareness can also help in following the threshold limit of noise in noise free zones. The question number 9 and 10 show a strong relationship between them. The figure also show that the respondents strongly agree with the effects of noise pollution, which means that most of them are suffering from hearing loss, depression and tension. Depression in turn causes lack of concentration in the work the same way as the sleeplessness and high blood pressure show close relationship.

The figure also highlight that the increase in population led to the increase in traffic density, which in turn is increasing noise pollution in the City. It also shows that the narrow roads and their bad condition is causing the noise pollution to increase. Government authorities like TEPA and EPA need to be active to Control the causes of noise.

5.8.2 Factor analysis

A questionnaire was developed to establish causes and effects of noise pollution. We identified many key factors of noise pollution and hearing loss. A very simple pattern of relationship among variables can be seen through factor analysis. It was concluded after the analysis that high traffic density, industries, unawareness about noise effects, complex and narrow road network and lack of traffic discipline are the major factor for the increase in noise pollution in the city. Another important variable, which is the hearing loss is supported by most of the factors in analysis.

The most common approach to decide the number of factors is a scree plot, which is a two dimensional graph with factors on the x-axix and eigenvalues on the y-axis. Eigenvalues represent the variance accounted for by each factor and these values are produced by principle component analysis. The first factor has eigenvalue of 6.3, which accounts for 27.7 % of the variance. It is clear from the scree plot that a few factors account for most of the variance, and then remaining factors have small eigenvalues. The sum of all eigenvalues is equal to the total number of variables. The component matrix and communalities are attached in the appendices.

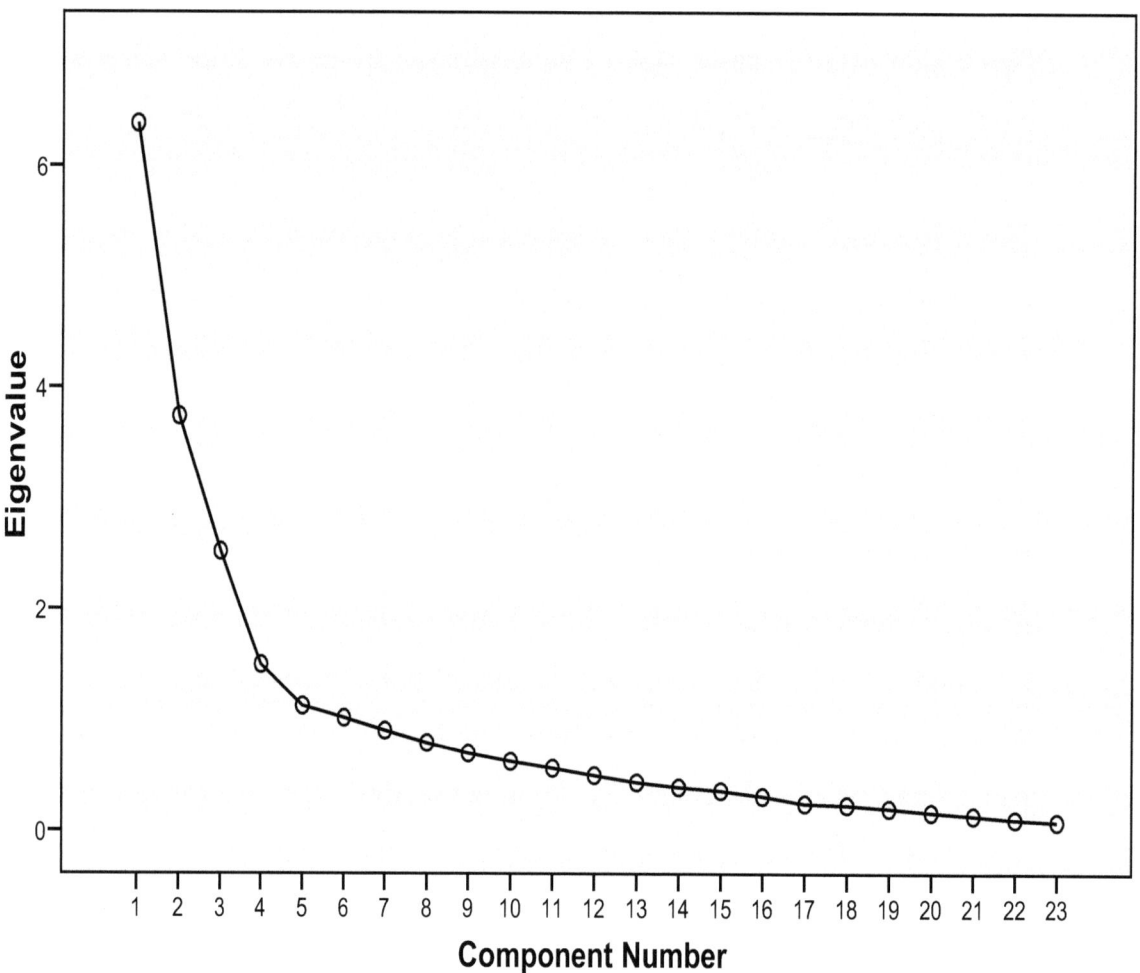

Figure 5.12: Scree plot

Table 5.8: Eigenvalues

Component	Initial Eigenvalues			Extraction Sums of Squared Loadings			Rotation Sums of Squared Loadings		
	Total	% of Variance	Cumulative %	Total	% of Variance	Cumulative %	Total	% of Variance	Cumulative %
1	6.383	27.751	27.751	6.383	27.751	27.751	5.099	22.170	22.170
2	3.741	16.264	44.015	3.741	16.264	44.015	4.215	18.325	40.495
3	2.519	10.952	54.967	2.519	10.952	54.967	2.445	10.631	51.126
4	1.498	6.512	61.478	1.498	6.512	61.478	1.882	8.181	59.306
5	1.125	4.890	66.368	1.125	4.890	66.368	1.354	5.888	65.195
6	1.017	4.420	70.788	1.017	4.420	70.788	1.287	5.594	70.788
7	.902	3.921	74.710						
8	.790	3.436	78.146						
9	.697	3.028	81.174						
10	.623	2.710	83.884						
11	.563	2.446	86.330						
12	.498	2.165	88.495						
13	.433	1.884	90.379						
14	.390	1.695	92.074						
15	.359	1.561	93.635						
16	.309	1.342	94.976						
17	.243	1.058	96.034						
18	.229	.996	97.031						
19	.199	.867	97.898						
20	.165	.716	98.614						
21	.136	.589	99.203						
22	.102	.444	99.647						
23	.081	.353	100.000						

Chapter 6:
Summary, conclusion, recommendation & future work

6.1 Summary

For any research conclusion is one of the most important part. In this section I reiterate the main contributions of the thesis. The objective of the thesis was to study the spatial patterns of noise pollution in Lahore city and to see the effects of noise pollution on human health especially on hearing.

Noise is a special feature of cities. It is a special kind of wave action usually transmitted by air in form of pressure waves and received by hearing apparatus present in body of animals. It is a sound without value that is undesired by the recipient. Noise is playing an ever increasing role in our lives and seems a regrettable but ultimately avoidable result of the current technology. Noise measurements include subjective as well as objective factors. A physical measurement of noise magnitude must be supported with subjective loudness and annoyance related factors. The sound pressure level is an objective quantification of noise based on measured sound pressure. The effect of noise on humans however, depends not only on its magnitude but also on its frequency content because the ear is not equally sensitive to noise at all frequencies in the audible range of 20-20,000 Hz. Noise levels in general have increased over the years. The noise levels in the cities have increased at about 1 dB per year for the last 30 years (Wang, 2005).

The data were collected from the sample sites. Data were collected from different sites using dBA. Measurements were kept as slow response and A weighted sound level was preferred for road traffic noise. The instrument read A grade weight values directly and the microphone was kept 1.2 meter above the ground at a distance of 1.5-5meter. The noise level was measured thrice a day. First readings were taken from 6 A.M. to 8 A.M. Second readings were taken from 12 P.M. to 2 P.M. Third readings were taken from 8P.M. to 10 P.M.

A preliminary survey was carried out among 125 inhabitants/workers by questionnaire method to gather information about the suffering of noise related health problems. The survey was conducted at those locations where the level of noise was high. These were eight sample points such as Badami Bagh inside, Taxali, Data Darbar, Sheranwala Gate, Infront of Shalimar Gardena, Railway Station, Garhi Shahu and Thokar Niaz Baig

The data collected by sound level meter and survey method was then presented in the form of maps, graphs, dendogram and scree plot.

A few questions were about causes of noise pollution. A few of them were about effects of noise pollution and rests of them were about suggestions for improvement.

The questionnaire method was used in the survey. Through questionnaire information was collected from the people about the causes and effects of the noise. Some physiological effects of noise were also studied during this research such as hearing loss, depression, tension, Insomnia and lack of concentration in the work.

Both the multivariate techniques cluster and factor analysis identifies that there is a strong relationship between two major groups of variables and variance between the factors successively.

6.2 Conclusion

Present study has given a fair idea about the noise pollution at different sample sites of the city. Noise levels were noted at seventy six sample sites, which were considered to be the most polluted sites. It was conclude that noise pollution has increased the maximum permissible limits of National Environmental Quality Standards at all the sample sites throughout the City. It showed that one hundred percent of all smple sites are exposed to higher noise levels.

The areas along walled city, Badami Bagh, Garhi Shahu, Railway Station, Jail road Canal Shadman Chowk and Chouburgi Chowk are the major locations of highest noise levels.

People living in these areas are directly suffering from diseases associated with noise. A marked difference between morning and evening noise levels has been noted. It has been identified that there is no decrease in noise pollution at some places and at some places where the noise levels decreases at evening time are exceeding the maximum permissible noise limits. More than 90 % people are suffering from hearing loss. Although this loss is temporary in the beginning as it is in the case of the respondents but may lead to permanent hearing loss.

Graduated symbols have been used to show the level of noise pollution. The smallest circle shows those areas where the noise level ranges between 47-70 dB (A). Next higher level of circle represents the areas where the noise level ranges from 71-80 dB (A). The medium sized circle represents areas where noise levels range from 81-90 dB (A). Next higher level of circle represents areas having noise level from 91-100 dB (A). The largest circle shows areas of highest level of noise pollution that is from 101-120. The graph along the map explains the noise values at different points of the city.

Badami Bagh was found to be the noisiest place and the cause of noise was unplanned vehicular traffic. Noise pollution at Yadgar Chowk, Taxali, outside Sheranwala Gate and Shah Alam Chowk falls between 91-100dB (A) and it exceeds the permissible limits. Yadgar Chowk, Ali Hajvery Darbar road and Taxali have heavy traffic flow but have wide roads due to which the load of traffic becomes less. At Sheranwala gate the volume of traffic is high and the road is narrow. Another important reason of noise in these areas is that the traffic entering the city passes through these points and then gets diverted to different roads. Shah Alam Chowk is another noisy place. It leads to Shah Alam Market, which is the Central Business District of the city. Ali Hajvery Darbar and Taxali chowk are also busy points where roads are although wide but load of traffic is very high. Mochi gate, Lohari, Urdu

Bazar, Gamey Shah and Do Morya bridge are the points where noise level ranges from 81-90 dB (A).

The noisiest places along Mall Road are the Secretariat Lahore and Chouburgi Chowk, Gwal Mandi and Lahore Hotel where noise levels were 97.6 dB (A) and 92.7 dB (A), 97.1 dB (A) and 91.3 dB (A) successively.

The noise level was also high at Shadman Chowk and at canal bank outside GOR-1. At Shadman chowk it was 95.1 dB (A) and 92.6 dB (A) successively. The major cause of high noise level at above discussed locations is high volume of traffic.

Railway Station is the most polluted sample site where noise level is 101.2 dB (A). The main cause of noise at Lahore Railway station is not only the railway but it is also a junction of intercity transport. Garhi Shahu is a midpoint between railway station and Canal Road. It is a busy chowk with many business activities and high traffic volume.

Buffers were also generated around the sample sites to see the extent and intensity of noise levels. The high intensity buffer areas have been shown in pink shade showing radius of 200 meters and the area surrounded by each buffer shows that the people living within this zone directly suffers from the noise related problems.

Then comes medium intensity buffers in yellow color are showing 150 meter radius. These buffers show noise level range from 71-90 dB (A). The cyan blue color of buffers shows the low intensities of noise. These buffers show noise level range with a radius of 50 meters.

Another analysis of changes in noise levels during morning and evening was done, which clearly can be seen on the maps.

6.3 Recommendations

There are many Recommendations that can bring improvement in the noise levels. These are as follows:

a. Development of an appropriate public transport system for the entire city to minimize the necessity of using private cars/vehicles. Recently some foreign (Daewoo) and local transporters have been allowed to run their buses through the city. This service should also be extended to various other areas of the city.

b. Introduction of flyovers/ additional underpasses can help in reduction of traffic load and in turn in noise pollution.

c. The system of railway needs to work better and massive transportation through railway within the City can be introduced because railway trains carry heavy traffic loads while comparatively creating lesser environmental pollution.

d. The area of one kilometer around hospitals and educational institutions can be made free from roads or use of horns can be strictly ban in these areas.

e. Noise free silencers need to be introduces in motor cycle rickshaws.

f. Traffic discipline need to be maintained and there can be some governmental checks on the traffic regulating authorities.

g. The use of ear protectors can be introduced and made compulsory in certain areas where the load of traffic cannot be reduced such as Bus Terminal and Railway Station.

h. Use of horns can be ban in residential areas.

i. Duration of work can be reduced to keep people safe from high noise levels.

j. Pressure horns can be removed from the public transport.

k. Traffic of the city can be managed properly by chalking out suitable plans.

l. Areas of potential improvements need to identify and a strategy can be framed at metropolitan level, covering all areas including road infrastructure development, maintenance management, public transport operation, associated infrastructure development, mass transit, traffic management and traffic control, parking traffic supervision and enforcement, environmental and social concerns, private sector participation, regulatory and financial aspects.

6.4 Scope for Future Work

a. In this work the noise levels were measured at seventy six sample sites, but more data can be collected to make a street level map and pattern of noise.

b. The effect of noise on animals and vegetation can be studied.

c. In this research the variables were only the level of noise and diseases but more variables such as speed of vehicles, material of road and presence of hindrances can also be measured.

d. In this research stress was given only on the vehicular noise but air traffic noise, industrial noise, noise produced by construction and many indoor sources can be studied further.

e. Development of road traffic noise model for Lahore City.

Annex-1: Bibliography

Integrated Master Plan for Lahore, (2021): Volume 1, National Engineering Services, Pakistan.

District Census Report of Lahore, (1998): December 2000 Population Census Organization, Government of Pakistan, Islamabad.

EPA, November (2006): "Measurement of noise level at different locations of Rawalpindi and Islamabad", Pakistan Environment Programme, EPA.

Pakistan Economic Survey, (2009-2010): Government of Pakistan, Finance Division, Advisor's Wing Islamabad, p. 242.

WHO, (1995): "Community Noise". http://www.noise.org/library/whonoise.htm/introduction. (Access date: 10-09-2010).

Stephenson, R. J., Vulkan, G.H. (1968): "Traffic noise", Journal of sound and vibration, vol. 7 (2), pp 247-262.

Agarwal, S., Swami, B.L. (2011): "Road traffic noise, annoyance and community health survey- A case study for an Indian city", Journal of noise and health, Jul-Aug; 13(53):272-6.

Thiery, L., Meyer, C. Bisch. (1988): "Hearing loss due to partly impulsive industrial noise exposure at levels between 87 and 90 dB (A), Journal of Acoustical Society of America, Vol. 84, Issue 2, pp. 651-659. http://asadl.org/jasa/resource/1/jasman/v84/i2/p651_s1 (accessed on 26-07-2011).

Basel, Birkhäuser. (1989): Journal "Sozial- und Präventivmedizin/Social and Preventive Medicine". u.o. : SpringerLink,.

Evans, G.W., Maxwell, L. (1997): Chronic noise exposure and reading deficits: The mediating effects of language acquisition. u.o.: Environment and Behavior.

Harman, D.M., Burgess, M.A. (1973): "Traffic noise in an urban situation", applied acoustics, vol. 6(4), pp 269-276.

Benedetto, G., Spagnolo, R.. (1977): "Traffic noise survey of Turin, Italy", applied acoustics, vol. 10 (3), pp 201-222.

Yeowart, N.S., Wilcox, D. J., Rossall, A.W: (1977) "Community reactions to noise from freely flowing traffic, motorway traffic and congested traffic flow", journal of sound and vibration, vol. 53 (1), pp 127-145.

Goswami, Sh.: (2009), "Road Traffic Noise: A case study of Balasore Town, Orissa, India", Int. J. Environmental Resources., 3(2):309-316, spring.

Banerjee, D. (2009): "Appraisal and mapping the spatial-temporal distribution of urban roadtraffic noise", Spring, Int. J. Environ. Sci. Tech., 6 (2), 325-335,.

Al-Ghonamy, A.I. (2010): "Analysis and evaluation of road traffic noise in Al-Dammam: A business city of eastern province of KSA", Journal of Environmental Science and Technology 3 (1): 47-55.

Nuzhat, Huma., Akhtar, Zahir, Shah. Mohammad., Qamar, Ismail. (1998): "Road traffic noise in Peshawar: An increasing problem", Department of community medicine, Khyber Medical College.

Raj, Kamla., Singh, Narendra. and Davar, S. C. (2004): "Noise Pollution- Sources, Effects and Control", J. Hum. Ecol., 16(3): 181-187 Haryana, India.

Roba, Mohammed., Saeed, Anis. (2010): "The Effects of Noise Pollution on Arterial Blood Pressure and Heart Pulse Rate of School Children at Jenin City", Palestine.

Ozer, Serkan., Yilmaz, Hasan., Yeşil, Murat., Yeşil, Pervin. (2009): "Evaluation of noise pollution caused by vehicles in the city of Tokat", Turkey, 2009. Scientific Research and Essays Vol. 4 (11), pp. 1205–1212, ISSN 1992- 2248 © Academic Journals. http://www.academicjournals.org/sre/abstracts/abstracts/abstracts%202009/Nov/Ozer%20et%20al.htm

Banerjee, D., Chakraborty, S. K., Bhattacharyya, S., Gangopadhyay, A. (2008): "Appraisal and mapping the spatial-temporal distribution of urban road traffic noise", West Bengal India. http://www.ceers.org/ijest/issues/full/v6/n2/602020.pdf

Singh, Lallan. Dr. (2010): "Environmental Geography", APH Publishing Corporation.

Craik, R. J., Stiring, M. J. R.. (1986): " Amplfied music as a nuisance", Appl. Acoust, p. 335-356.

Smith, B.J. Peter, R. J., Owen, S. (1996): "Acoustics and noise control", 2nd Edition, Published by addison Wesley Longman Limited, p.1.

Farcaş, Florentina. (2008): "Road Traffic Noise: A study of Skåne region", Sweden. http://www.isprs.org/proceedings/XXXVIII/2-W11/Farcas_Sivertun.pdf

Mehdi, Muhammad. Raza. (2002): "Spotting noise risk zones in Karachi Pakistan : A GIS Perspective", Karachi. http://www.gisdevelopment.net/proceedings/gisdeco/sessions/Mehdipf.htm

Harabidis, A. S. (2008): "Acute Effect of Night-Time Noise Exposure on Blood Pressure in Population Living Near Airports", European Heart Journal. http://www.najah.edu/thesis/5171651.pdf

Harris, C.M. (1979): "Handbook of noise control", 2nd Edition, Mc Graw Hill, New York, p.148-149.

Verwijmeren, M. A. P. (1987): "An information for the introduction of new occupational noise legislation in the Netherland", , Inter-noise 87, p.1613-1616.

Lord, P., Thomas, F. L. (1963): "Noise measurement and control", Heywood and Company Ltd, London, p. 12-62.

Hay, B. (1981): "Maximum permissible noise levels in the work place in the EEC, Spain, Portugal and Turkey", Appl. Accoust, p. 61-69.

Lehman, G., J, Tamm. (1956): "Changes of circulatory dynamics of resting men under the effects of noise", Intl. Z Angew Physiolo., p.217-227.

Katyal, Timmy. Satake, M. (1989): "Environmental Pollution", Anmol publications, New Delhi, p.219.

Kupchella, Charles. E., Hyland, Margaret. C. (1993): "Environmental Science: Living within the system of nature", Third Edition, Prentice Hall, Inc., p.438.

Kraus, S.K., Ding, D., Jiang, H., Lobarinas, E., Sun, W., Salvi, R. J. (2011): "Relationship between noise-induced hearing-loss, persistent tinnitus and growth-associated protein 43 expression in the rat cochlear nucleus: does synaptic plasticity in ventral cochlear nucleus suppress tinnitus", Neuroscience Jul 28.

http://www.unboundmedicine.com/medline/ebm/record/21821100/full_citation/Relationship_between_noise_induced_hearing_loss_persistent_tinnitus_and_growth_associated_protein_43_expression_in_the_rat_cochlear_nucleus:_does_synaptic_plasticity_in_ventral_cochlear_nucleus_suppress_tinnitus (accessed on 11-08-2011).

Carmen, Richard., Uram, Shelley. (2002): "The Hearing loss anxiety in adults", The hearing journal, April -Vol 55-Issue 4-pp48, 50, 52-54.

http://journals.lww.com/thehearingjournal/Fulltext/2002/04000/Hearing_loss_and_anxiety_in_adults.6.aspx

Singal, SP. (2005): "Noise Pollution and Control Strategy", Narosa publishing house, New Delhi India, pp 103-106.

Berglund, B., Lindvall, T. (1995): "Community noise", Archives of the Center for Sensory Research, Printed by Jannes Snabbtryck, Stockholm, Sweden, 2(1), 1-195.

Downey, Micheal. (2003): "Deaf Defining: Today's noise pollution solutions", Aug. http://findarticles.com/p/articles/mi_m0FKA/is_8_65/ai_105370810/ (accessed on 12-08-2011).

Kurakula, Vinaykumar. (2007): "A GIS_Based Approach for 3 D Noise Modelling Using 3D City Model". http://www.gem-msc.org/Academic%20Output/Kurakula%20Vinay.pdf (accessed on March 2011).

Stephen, A., Stansfeld, Mark, P., (2007): Matheson, "Noise pollution: non-auditory effects on health", British Medical Bulletin, Volume 68, Issue 1, pp 243-257. http://bmb.oxfordjournals.org/content/68/1/243.abstract (Accessed on 14-08-2011).

Lisa, Goines. R. N., Louis, Hagler, M.D. (2007): "Noise Pollution: A Modern Plague", Southern Medical Journal, Lippincott Williams and Wilkins, pp 287-294. http://www.medscape.com/viewarticle/554566 (Accessed on 14-08-2011).

Suh, Meei. Hsu., Wen, Je. Ko. (2010): "Associations of exposure to noise with physiological and psychological outcomes among post cardiac surgery patients in ICUs", Clinics (Sao Paulo), October, pp 985-989.

Dalton, D.S., Cruickshank, K.J., Wiley, T. L., Klein, B. E., Klein, R., Tweed, T.S. (2001): "Association of leisure-time noise exposure and hearing loss", Jan-Feb, 40(1), pp 1-9.

Richard, Neitzel., Noah, S., Seixas, Janic. Camp., Michael, Yost. (1999): "An Assessment of Occupational Noise Exposures in Four Construction Trades", American Industrial Hygiene Association Journal, 60(1), 807-817. http://www.ncbi.nlm.nih.gov/pubmed/10635548 (accessed on 17-08-2011).

Davies, H. W., Teschke, K., Kennedy, S. M., Hodgson, M. R., Demers, P. A. (2009): "A retrospective assessment of occupational noise exposures for a longitudinal epidemiological study", Occup Environ Med, Volume 66, Issue 6, pp 388-394.

Wang, Qi. (2008): "Application of grey relational analysis and hierarchical cluster analysis in regional environmental noise pollution", IEEE Xplore, Issue Date 18-20 Nov. 2007,pp 336 – 340, Nanjing. http://ieeexplore.ieee.org/xpl/freeabs_all.jsp?arnumber=4443292: http://www.env.go.jp/en/air/noise/noise.html: http://www.osha.gov/pls/oshaweb/owadisp.show_document?p_table=standards&p_id=9735

Alam, Ahmad. Rafay. (2008): "Facing urban congestion", The News, July 3.

http://lahorenama.wordpress.com/2008/07/03/facing-urban-congestion/(date of access 1-09-2011).

Wang, Lawrence. K., Pereira, Norman. C., Hung, Yung-Tse. (2005): "Advanced air and noise pollution", Human Press, Totowa, New Jersey.

Annex-2:
NEQS for noise pollution

National Environmental Quality Standards for Noise

S. No.	Category of Area / Zone	Effective from 1st July, 2010		Effective from 1st July, 2012	
		Limit in dB(A) Leq *			
		Day Time	Night Time	Day Time	Night Time
1.	Residential area (A)	65	50	55	45
2.	Commercial area (B)	70	60	65	55
3.	Industrial area (C)	80	75	75	65
4.	Silence Zone (D)	55	45	50	45

Note:

1. Day time hours: 6.00 a.m to 10.00 p.m.

2. Night time hours: 10.00 p.m. to 6.00 a.m.

3. Silence zone: Zones which are declared as such by the competent authority. An area comprising not less than 100 meters around hospitals, educational institutions and courts.

4. Mixed categories of areas may be declared as one of the four above-mentioned categories by the competent authority.

*dB(A) Leq: Time weighted average of the level of sound in decibels on scale A which is relatable to human hearing.

[No. F. 1(12)/2010-11-General.]

MUHAMMAD KHALIL AWAN,
Section Officer (PEPC).

Annex-3:
Questionnaire: Causes and effects of noise pollution

(Title)

Date: _____   Place: _____

Age: _____   Gender: _____

Period of residence: _____ Exposure to noise _____

Are you or any of your family members suffering from hearing loss?   Yes------- No---------

If yes how long you have been suffering from this problem---?

Q.1- Please read the statements carefully and mark the option that is suitable according to you:

Statement	Strongly Agree	Agree	Neutral/ Not decided	Disagree	Strongly Disagree
1- Traffic density is the root-cause of increasing noise pollution.					
2- Population is also one of the major causes of increase in number of vehicles and noise pollution.					
3- Penetration of industries/ factories in the non-commercial areas is another cause of noise pollution.					
4- Business activities are the cause of noise pollution at this place.					
5- Illiteracy/ unawareness regarding traffic rules and regulations is making its part for violation of traffic laws that ultimately is contributing to noise pollution.					
6- Bad condition of roads are causing noise pollution					
7- Lack of Traffic discipline causes more noise pollution					
8- Complex and narrow road network are the reasons for increasing traffic density.					
9- Diseases like frustration and stress are mainly due to noise pollution.					
10- Are you suffering from hearing loss?					
11- It is cause of depression and tension for you					

12- It is cause of high blood pressure for you					
13- It is cause of sleeplessness for you					
14- It is cause of lack of concentration for you					
15- A mature road network is necessary in place of existing road network.					
16- Government authorities like TEPA, EPA are playing their role in controlling noise pollution.					
17- If only the noise produced by large, heavy vehicles/ buses is controlled, noise pollution can be reduced to an acceptable level.					
18- Self-learning and awareness regarding noise pollution is as important as physical environment friendly network of roads and industries.					
19- Noise pollution awareness campaigns should be introduced by the government to address and reduce noise pollution.					
20- Strict actions should be taken for implementing traffic rules as well as environmental protection laws.					
21- Hospitals and educational places like universities, colleges and schools should be declared as "Noise-free" zones.					
22- A threshold limit of noise level should be fixed around these places in a specific buffer zone by the law enforcing authorities.					

Annex-4:
Noise levels at seventy six sample sites

Id	Location	Morning	Noon	Evening	Sound Level (dB)
1	A.G.Office (1)	85	95.2	88	89.4
2	Ali Town Raiwind Road (2)	82	84.5	80	82.2
3	Anarkali (3)	79.5	83	78	80.2
4	Assembly hall (4)	73	78	79	76.7
5	Badami Bagh (Bus Terminal In) (5)	115	125	121.3	120.4
6	Badami Bagh (Bus Terminal Out) (6)	103	110	89	100.7
7	Bhatti Chowk (7)	87	98.3	88	91.1
8	Bohr Wala Chowk (8)	87	93.8	83	87.9
9	Campus Bridge (9)	80	80.6	76	78.9
10	Canal Bank Road (Outside GOR-1) (10)	88.4	91.3	98	92.6
11	Chah Blaqi Wala (11)	48	49.2	44	47.1
12	Chuburji Chowk (12)	90	95.8	92.2	92.7
13	Club Chowk (13)	76	81.3	80	79.1
14	CMA Colony Cantt (14)	60	71	66	65.7
15	Data Darbar (15)	88	94.8	90	90.9
16	Dharampura Bridge (16)	70.6	88.2	68.1	75.6
17	Do-Morya Bridge (17)	87	91.8	84	87.6
18	Doctors Hospital (18)	86	85	83	84.7
19	Doctors Hospital (Underpass) (19)	95	102	99	98.7
20	Dubai town (By-pass) (20)	87	81.3	75	81.1
21	Eden Villas (21)	60	61.5	59	60.2
22	Ferozpur Road (Bus Stop) (22)	85	81.3	79.7	82.0
23	Fortress Stadium (Outside) (23)	70	75	81	75.3
24	Gamey Shah (24)	75	86	80	80.3
25	Garhi Shahu 1 (25)	50	66	60	58.7
26	Gari Shahu 2 (26)	55	57	51	54.3
27	Garhi Shahu Bridge (27)	99	96.3	89	94.8
28	GOR 1 (28)	65	70.8	67	67.6
29	GOR 1 (At Distance) (29)	57	62.2	52	57.1
30	Government Saleem Model School (30)	56	66	70	64.0
31	Governor House (31)	88.5	91.3	82.6	87.5
32	GPO Chowk (32)	88	89.3	83.5	86.9
33	Gulberg Canal (33)	100	112	98	103.3
34	Gwal Mandi (34)	100	102.4	89	97.1
35	Hall Road (35)	86	92.3	81.7	86.7
36	High Court (36)	67	77	71	71.7
37	Hussain Chowk (37)	77	84.8	80.3	80.7
38	Imam Bargah (38)	65	70.7	67	67.6

39	Jail Road Canal (39)	82	87.4	77	82.1
40	Jain Mandar (40)	80	90.3	90	86.8
41	Jinnah Hospital (41)	85	84.3	85	84.8
42	Kalma Chowk (42)	80	86.8	89	85.3
43	Lahore College (43)	95	108.4	99.2	100.9
44	Lahore Hotel (44)	90	90.8	93	91.3
45	Lane 1 Railway Station (45)	70	79.1	78	75.7
46	Lane 2 Railway Station (46)	67	72	70	69.7
47	Liberty Chowk (47)	83	85.3	89	85.8
48	Lohari (48)	89	92.3	84	88.4
49	M.A.O. College (49)	80	87.6	81	82.9
50	Mall Road Canal (50)	85	87.8	85	85.9
51	Mayo Garden 1 (51)	59	60.2	58	59.1
52	Mayo Garden 2 (52)	53	55.3	45	51.1
53	Mayo Garden (Outside) (53)	85	90.3	88	87.8
54	Mazzang Chungi (54)	88	95	90	91.0
55	Minhaj College Township (55)	90	88	74.4	84.1
56	Mochi Gate (56)	90	94.1	77	87.0
57	Model Town Park (57)	65	70	67	67.3
58	Muslim League House (58)	90	89.9	78	86.0
59	Neela Gumband (59)	77	81.8	76	78.3
60	Old Campus (60)	80.5	84.8	77	80.8
61	Railway Station (61)	98	117.4	88.1	101.2
62	Regal Chowk (62)	70	84	80	78.0
63	Safan Wala Chowk (63)	80	92.3	90	87.4
64	Secretariat (64)	95	100.8	97	97.6
65	Shadman Chowk (65)	99	96.3	90	95.1
66	Sha-Alam Chowk (66)	85	98.6	100	94.5
67	Shalimar Garden (67)	97.5	90	87	91.5
68	Sheranwala Gate (Inside) (68)	66	68.1	65	66.4
69	Sheranwala Gate (Outside) (69)	85	93.3	98	92.1
70	Simla Hill (70)	80	87.3	87	84.8
71	Sir Ganga Ram Hospital (71)	83	85	88.3	85.4
72	Taxali (72)	99	98.8	87	94.9
73	Thokar Niaz Baig (73)	80	86.3	99.4	88.6
74	Urdu Bazar (74)	80	94	72	82.0
75	Yadgar Chowk (75)	102	101.3	88	97.1
76	Zafar Shaheed Chowk (76)	99	102.9	88	96.6

Annex-5:
Communalities

	Initial	Extraction
Traffic density is the root -cause of increasing noise pollution.	1.000	.610
Population is also one of the major causes of noise pollution.	1.000	.788
Penetration of industries/factories in the non-commercial areas is another cause of noise pollution.	1.000	.758
Busniess activities are the cause of noise pollution at this place.	1.000	.600
Illiteracy/unawareness regarding traffic rules and regulations is making its part for violation of traffic laws that ultimately is contributing to noise pollution.	1.000	.708
Bad condition of roads are causing noise pollution.	1.000	.760
Lack of traffic discipline causes more noise pollution.	1.000	.701
Complex and narrow road network are the reasons for increasing traffic density.	1.000	.721
Diseasis like hearing loss, frustration,stresss,depressionetc are mainly due to noise pollution.	1.000	.849
Noise is the major cause of hearing loss.	1.000	.856
It is cause of depression and tension.	1.000	.856
It is cause of high blood presure.	1.000	.804
It is cause of sleeplessness.	1.000	.807
It is cause of lack of concentration.	1.000	.827
A mature road network is necessary in place of existing road network.	1.000	.678

Government authorities like TEPA,EPA are playing their role in controlling noise polution.	1.000	.766
If only the noise pollution is produced by large,heavy vehicles/buses is controlled, noise pollution can be reduced to an acceptable leval.	1.000	.680
self-learning and awareness regarding noise pollution is as important as physical environment friendly netwok of roads and industries.	1.000	.479
Noise pollution awareness compaigns should be introduced by the government to address and reduce noise pollution.	1.000	.595
Strict actions should taken for implementing traffic rules as well as environmental protection laws.	1.000	.645
Hospitals and educational places like universities,colleges and schools should be declared as " Noise-free" zones.	1.000	.648
A threshold limit of noise level should be fixed around these places in a specific buffer zone by the law enforcing authorities.	1.000	.583
Use of horn at the time of traffic jam should be reduced.	1.000	.562

Annex-6:
Component matrix(a)

	Component					
	1	2	3	4	5	6
Traffic density is the root-cause of increasing noise pollution.	-.265	.421	.153	.479	.166	-.288
Population is also one of the major causes of noise pollution.	-.195	.395	-.242	.644	-.157	.309
Penetration of industries/factories in the non-commercial areas is another cause of noise pollution.	-.355	.718	-.200	.178	-.071	.201
Busniess activities are the cause of noise pollution at this place.	.531	-.278	-.210	.178	-.174	.367
Illiteracy/unawareness regarding traffic rules and regulations is making its part for violation of traffic laws that ultimately is contributing to noise pollution.	-.026	.288	.185	.220	.716	-.168
Bad condition of roads are causing noise pollution.	.631	-.501	-.054	.113	.260	.171
Lack of traffic discipline causes more noise pollution.	.707	-.397	-.159	-.013	.116	.067
Complex and narrow road network are the reasons for increasing traffic density.	.614	-.523	-.066	.037	.245	-.062
Diseasis like hearing loss, frustration,stresss,depressionetc are mainly due to noise pollution.	.779	.420	.110	-.126	-.075	-.180
Noise is the major cause of hearing loss.	.804	.418	.139	.041	-.029	-.115
It is cause of depression and tension.	.774	.492	.112	-.031	-.032	.032
It is cause of high blood presure.	.698	.519	.080	-.117	.009	.164
It is cause of sleeplessness.	.699	.481	.164	-.218	-.097	-.057

It is cause of lack of concentration.	.796	.418	.119	-.058	.004	.017
A mature road network is necessary in place of existing road network.	.695	-.357	-.186	.106	-.096	.109
Government authorities like TEPA,EPA are playing their role in controlling noise polution.	.473	-.300	-.091	.471	-.261	-.393
If only the noise pollution is produced by large,heavy vehicles/buses is controlled, noise pollution can be reduced to an acceptable leval.	.548	-.310	-.153	.491	.133	.039
self-learning and awareness regarding noise pollution is as important as physical environment friendly netwok of roads and industries.	.063	-.012	.543	-.002	.334	.262
Noise pollution awareness compaigns should be introduced by the government to address and reduce noise pollution.	.197	-.160	.537	.169	-.320	-.333
Strict actions should taken for implementing traffic rules as well as environmental protection laws.	.071	-.275	.702	-.100	-.027	.247
Hospitals and educational places like universities,colleges and schools should be declared as " Noise-free" zones.	-.088	-.325	.659	.107	-.234	.183
A threshold limit of noise level should be fixed around these places in a specific buffer zone by the law enforcing authorities.	-.218	-.346	.620	.122	-.013	-.127
Use of horn at the time of traffic jam should be reduced.	-.135	.377	.483	.324	-.035	.250

Extraction Method: Principal Component Analysis.
a 6 components extracted.

Annex-7:
Rotated Component Matrix(a)

	Component					
	1	2	3	4	5	6
Traffic density is the root-cause of increasing noise pollution.	-.010	-.314	-.029	.387	.528	.287
Population is also one of the major causes of noise pollution.	-.062	-.041	-.179	.866	.031	.014
Penetration of industries/factories in the non-commercial areas is another cause of noise pollution.	.098	-.504	-.278	.607	.076	-.207
Busniess activities are the cause of noise pollution at this place.	.168	.652	-.003	.162	-.344	-.052
Illiteracy/unawareness regarding traffic rules and regulations is making its part for violation of traffic laws that ultimately is contributing to noise pollution.	.094	-.037	.026	.042	.826	-.115
Bad condition of roads are causing noise pollution.	.117	.833	.109	-.188	.035	-.065
Lack of traffic discipline causes more noise pollution.	.254	.748	-.046	-.256	-.094	-.003
Complex and narrow road network are the reasons for increasing traffic density.	.109	.757	.028	-.352	.068	.078
Diseasis like hearing loss, frustration,stresss,depressionetc are mainly due to noise pollution.	.891	.123	-.071	-.107	.018	.153
Noise is the major cause of hearing loss.	.879	.215	-.019	.029	.093	.166
It is cause of depression and tension.	.905	.168	-.019	.081	.032	.009
It is cause of high blood presure.	.868	.121	-.020	.093	-5.48E-005	-.162
It is cause of sleeplessness.	.893	.017	-.002	-.083	-.042	.023

It is cause of lack of concentration.	.882	.218	-.003	.012	.042	.006
A mature road network is necessary in place of existing road network.	.259	.730	-.051	-.083	-.241	.103
Government authorities like TEPA,EPA are playing their role in controlling noise polution.	.103	.510	-.102	.041	-.051	.693
If only the noise pollution is produced by large,heavy vehicles/buses is controlled, noise pollution can be reduced to an acceptable leval.	.097	.765	-.042	.160	.124	.206
self-learning and awareness regarding noise pollution is as important as physical environment friendly netwok of roads and industries.	.108	.053	.565	-.020	.290	-.249
Noise pollution awareness compaigns should be introduced by the government to address and reduce noise pollution.	.155	.010	.450	-.106	-.050	.595
Strict actions should taken for implementing traffic rules as well as environmental protection laws.	.033	.061	.782	-.148	-.067	-.054
Hospitals and educational places like universities,colleges and schools should be declared as " Noise-free" zones.	-.143	.012	.758	.025	-.160	.162
A threshold limit of noise level should be fixed around these places in a specific buffer zone by the law enforcing authorities.	-.275	-.067	.624	-.138	.127	.279
Use of horn at the time of traffic jam should be reduced.	.155	-.254	.449	.490	.179	-.013

Extraction Method: Principal Component Analysis.
Rotation Method: Varimax with Kaiser Normalization.
A Rotation converged in 6 iterations.

Annex-8:
Component Transformation Matrix

Component	1	2	3	4	5	6
1	.751	.632	-.016	-.144	-.066	.106
2	.602	-.580	-.243	.416	.221	-.139
3	.186	-.227	.917	-.097	.191	.162
4	-.178	.341	.052	.739	.325	.443
5	-.082	.211	-.047	-.210	.835	-.453
6	-.021	.227	.307	.454	-.327	-.736

Extraction Method: Principal Component Analysis.
Rotation Method: Varimax with Kaiser Normalization.